PRINCIPLES OF THERMODYNAMICS AND STATISTICAL MECHANICS

PRINCIPLES OF THERMODYNAMICS AND STATISTICAL MECHANICS

D. F. Lawden

University of Aston in Birmingham

JOHN WILEY & SONS

Chichester · New York · Brisbane · Toronto · Singapore

Copyright © 1987 by John Wiley & Sons Ltd.

All rights reserved.

No part of this book may be reproduced by any means, or transmitted, or translated into a machine language without the written permission of the publisher.

Library of Congress Cataloging-in-Publication Data:

Lawden, Derek F.
  Principles of thermodynamics and statistical
mechanics.

  Bibliography: p.
  Includes index.
  1. Thermodynamics.   2. Statistical mechanics.
  I. Title.
QC311.L34   1986   536′.7   86-13324

ISBN 0 471 91172 0
ISBN 0 471 91174 7 (pbk.)

British Library Cataloguing in Publication Data:

Lawden, D. F.
  Principles of thermodynamics and
  statistical mechanics.
  1. Statistical thermodynamics
  I. Title
  536′.7   QC311.5

ISBN 0 471 91172 0
ISBN 0 471 91174 7 Pbk

Printed and bound in Great Britain

Contents

Preface

It is, perhaps, not surprising that the mathematical analysis of the behaviour of physical systems which comprise only a small number of elements, whose interactions are subject to physical principles which are themselves expressed in mathematical form, is so strikingly successful. The discovery that elementary particles and their fields are governed by simple, universal laws revealed a truly remarkable feature of the world, but given this circumstance, the relevance of mathematics as an instrument for predicting the behaviour of simple systems was clearly indicated. As would be expected, therefore, the early triumphs of mathematical physics consisted in explaining the behaviour of systems such as the pendulum and the sun with its attendant planets, which could be modelled using a small number of elements. However, one of the more astonishing achievements of natural philosophers in the nineteenth century was to demonstrate that complex systems, comprising a myriad of particles in a state of total disorder, were also amenable to mathematical treatment and that distinctive derivative physical laws governing all such systems could be distilled from the more fundamental principles applicable to their individual molecules. Thus Maxwell, Boltzmann and Gibbs succeeded in showing, for the first time, how law and order could arise in a natural manner out of chaos.

This is the role of statistical mechanics, viz. to reinterpret the primary physical laws in the light of the chaos normally present in those macroscopic systems which are the common objects of experimental observation. It is accordingly a subsidiary theory, but since a very common feature of physical phenomena is the disorder reigning amongst the responsible elements, it is a theory having an exceptionally wide field of application and is correspondingly more often used by practising physicists than any other. Another consequence of the secondary status of the theory is that its methods are largely independent of the precise nature of the primary physical elements and the laws which constrain them; these methods have been devised to cope with disorder and this dominates the form they take. Thus statistical mechanics was easily able to accommodate the revolutionary upheaval in fundamental particle theory which occurred in the 1920s when Newtonian laws were replaced by the principles of quantum mechanics; the necessary amendments were made without delay and, indeed, some had already been tentatively introduced (e.g. by Planck) to rectify anomalies appearing in the classical statistical mechanics, thus initiating the search for a new theory of mechanics. When, in their turn, the quantum principles become unacceptable and a more comprehensive theory is formulated, it is to be expected that statistical mechanics will retain its present structure, requiring no more than relatively minor amendments. Indeed, since the quantum mechanics appears to be a perfectly adequate instrument for predicting the motions of atoms and molecules under the influence of their mutual interactions, and it is to such systems of elements that statistical mechanics has thus far been almost exclusively applied, a more comprehensive theory of mechanics (which must include quantum mechanics as a limiting case) is unlikely to disturb the statistical theory as it stands at present, but will only permit its extension to more sophisticated systems.

A consequence of the wide field of application of the statistical theory is that the number of available textbooks is very large. However, the great majority of these have been written by physicists for their students and emphasize results obtained when the theory is applied to specific systems of particular interest to the experimentalist. Many of these books pay scant regard to the principles upon which the theory is based and to the logical development of the argument erected upon these foundations—the same slipshod reasoning is often found repeated in text after text as the students of yesteryear become the authors of today and regurgitate errors they have been conditioned to accept. My aim, therefore, has been to provide a clear understanding of the general principles upon which the theory is founded, to render these ideas plausible and then to develop mathematically the main lines of argument leading to general results which can be applied to a multitude of physical situations, only a few of the more important of these being illustrated in the text. Thus any reader who completes this book should be well equipped to apply the theory to any specific physical phenomenon of a relevant type in which he may be chiefly interested and can consult a work devoted to the details of this application, confident that his background knowledge will be sufficient to permit his understanding the contents.

The level of both pure and applied mathematical expertise assumed to be possessed by the reader is necessarily quite high, but should be within the competence of a third-year university student following a course in mathematical physics. Key pure mathematical techniques used are: partial differentiation (e.g. the 'chain rule') and perfect differentials; multiple integrals, including change of variables; extrema of a function of several variables subject to constraints; contour integration and calculus of residues. In regard to his grounding in applied mathematics, the reader should be familiar with classical analytical mechanics to Hamilton's equations, and the various statistical measures of a distribution, like the mean and variance. He should also have read physics to at least A-level and have studied (or be concurrently studying) non-relativistic quantum mechanics; an acquaintance with Maxwell's theory of the electromagnetic field will be helpful in a few sections.

The book is planned along chronological lines. Thus the first two chapters contain the main elements and results of the classical theory of thermodynamics; these endeavour to place the reader in an appropriate frame of mind to appreciate the deeper understanding of thermodynamical phenomena attainable on the basis that heat is the energy of random motion of the molecules of a body. This idea is introduced in Chapter 3 where, using the principle of randomicity in phase and the Gibbs microcanonical ensemble, the classical results for a gas, first derived by Maxwell and Boltzmann on a basis of Newtonian mechanics, are established. No apology need be offered for giving so much space to a theory which was later to be superseded by one based on quantum principles since, as has been remarked, the characteristic ideas of statistical mechanics are largely independent of the fundamental principles of mechanics we accept and didactically it is therefore judicious first to present these seminal ideas in as simple a context as can be found. Further, every student of the subject should be made aware of the anomalies inherent in the classical theory which were removed by the change to quantum principles. The fundamental method of mean values, due to Darwin and Fowler, is described in Chapter 4 and applied first to an isolated gas, whose energy is presumed given, and later to an isolated system comprising a heat bath in which an arbitrary system is immersed—this leads to the idea of the canonical ensemble. In this chapter, some applications of the theory to diatomic gases, paramagnetic materials and crystals are made. In Chapter 5, the quantum theory for a Gibbs ensemble is developed, replacing the classical phase density function by a density matrix and the principle of randomicity in phase by one of randomicity of eigenstates. The theory is further generalized to apply to ensembles representing open systems which exchange particles as well as heat with the environment; the grand canonical ensemble is introduced to model such systems. Chapter 6 then applies the quantized theory to crystals and paramagnetic materials, and the final chapter makes use of the grand canonical ensemble to derive the Bose–Einstein and Fermi–Dirac statistics for ideal gases of bosons and fermions respectively; a section is also devoted to the analysis of radiation regarded as a photon gas. There are three mathematical appendices and a bibliography, and each chapter is accompanied by a set of exercises.

The book is based upon a course of lectures I gave to third-year mathematics students over a number of years at the University of Aston in Birmingham and most sections have therefore been class-tested. In my experience, students find this aspect of mathematical physics more intellectually challenging than any other and it is important, therefore, that the teacher should plan his presentation of the topic very carefully and make every effort to be thoroughly lucid. This has been my chief intention in writing this account.

D. F. LAWDEN

List of constants

In the SI system of units:

Avogadro's number	$= L = 6.0221 \times 10^{23} \, \text{mol}^{-1}$
Bohr magneton	$= \mu_B = 9.27406 \times 10^{-24} \, \text{J T}^{-1}$
Boltzmann's constant	$= k = 1.38066 \times 10^{-23} \, \text{J K}^{-1}$
Gas constant	$= R = 8.3145 \, \text{J mol}^{-1} \text{K}^{-1}$
Planck's constant	$= h = 6.6261 \times 10^{-34} \, \text{Js}$
Velocity of light	$= c = 2.998 \times 10^{8} \, \text{m s}^{-1}$

CHAPTER 1

Classical thermodynamics

1.1 Thermodynamics and statistical mechanics

Thermodynamics is the branch of physics which studies the manner in which the properties of substances are dependent upon their temperatures. The classical theory was developed during the nineteenth century and succeeded in identifying the quantities needed to describe the thermodynamic state of a body (e.g. temperature, pressure, entropy, etc.) and in predicting the manner in which these will vary when substances in different states interact by mixing or by being placed in thermal contact with one another. Coefficients (e.g. specific heat) were defined whose values determined the thermal properties of a substance, but it was necessary to determine these values empirically—the theory provided no means by which they could be deduced from other physical properties of the substance, such as its crystalline structure. In particular, the theory took no account of the molecular constitution of substances and the mechanical or electromagnetic properties of the particles from which they are composed. The possibility that the thermodynamic behaviour of matter might be explained as a consequence of the mechanical behaviour of its constituent molecules, was first successfully exploited by Clerk Maxwell (1831–79) and later formed the basis for a new theory called *statistical mechanics*, which was created by Ludwig Boltzmann (1844–1906). Thus the principles of classical thermodynamics have been shown to be derivable from the more fundamental principles of mechanics and electromagnetism. Nevertheless, the derived principles remain of prime importance for physics and it is still sensible to study them in isolation, before exhibiting them as elements of a more comprehensive, and thus necessarily more complex, theory. By this means, they will be brought to the centre of the reader's attention and the mathematical relationships connecting them will be emphasized.

In this and the following chapter, the elements of the classical theory will accordingly be constructed upon a foundation of experimental facts relating to the macroscopic properties of materials alone and no reference will be made to microscopic properties associated with their molecular structures. In subsequent chapters, we will demonstrate how these principles arise from a consideration of the mechanical behaviour of atoms and molecules and how the thermal characteristics of a substance can be predicted from a knowledge of its microscopic structure.

1.2 Thermodynamical equilibrium

The theory we are about to develop applies to the widest possible variety of physical systems; these may be homogeneous or heterogeneous and include solids, liquids and gases (and even radiation) which are mixed together or separated from one another in containers. However, it will almost always be assumed that the system under study has arrived at a state of equilibrium, in the sense that the macroscopic properties of its constituents are not observed to change as further time elapses. The properties here referred to are those such as density, pressure, temperature, magnetization, etc. which can be measured by instruments which do not probe the microscopic structure of the system. It is now well understood that this *thermodynamical equilibrium*, as it is termed, is superficial and is compatible with a rapid variation in the state of the system at the microscopic level of its constituent molecules. Thus if a system comprising a colloidal suspension of solid particles in a liquid is allowed to reach thermodynamic equilibrium in a vessel kept at a uniform temperature, a microscopic examination of the particles reveals that they are propelled into random motions by the bombardment they receive from surrounding liquid molecules (*Brownian motion*) and thus that the equilibrium at the macroscopic level is of a statistical nature and is not absolute. Such microscopic fluctuations are ignored by the classical theory, but can be treated by the more fundamental methods of statistical mechanics.

A system may approach equilibrium either by being placed in a container which shields it from all external influences (e.g. a Dewar's flask), or by being permitted to interact with a steady environment through the walls of its container. In the former circumstances, the system is said to be *isolated* and the walls of the container to be *adiabatic*. In the latter case, the system is said to be in *thermal contact* with its surroundings and the walls to be *diathermal*. We shall often be supposing that the temperature of the system is known and this will imply that it has been placed in thermal contact with an environment at this temperature or, as we shall say, has been *immersed in a heat bath*. However, up to this point, no precise meaning has been given to the term 'temperature' and it will be our immediate object in the next section to remedy this deficiency.

We shall often be supposing that a system is transferred between a pair of terminal equilibrium states via a succession of intermediate non-equilibrium states. This will be termed a *process*. If, however, the transfer takes place so slowly that the system is virtually in a state of equilibrium throughout, the process will be described as being *quasi-static*.

1.3 Zeroth law of thermodynamics

Suppose a system has been permitted to settle into a state of thermodynamical equilibrium. To distinguish this state from all other possible equilibrium states of the system, it will be necessary to measure the steady state values of various macroscopic physical quantities associated with it. For a given system, we can then identify a minimal set of these quantities, whose values being known, the values of all other such quantities are found to be determined. This is termed a complete set of *parameters* or *variables of state* for the system. In the particular case of a homogeneous gas, its pressure P and volume V are found to constitute a complete set of parameters of state.

Next suppose that a pair of systems A and B have arrived at equilibrium states independently of one another. It is found that, in general, if they are isolated and then placed in thermal contact with one another, they will not remain in equilibrium. The reader will, of course, be aware that the condition that the combined system A + B should be in equilibrium is that the temperatures of A and B are the same. We can accordingly base a precise definition of the *temperature of a system* upon this fact of observation.

But it is first necessary to state an associated principle, which is also confirmed by experiment. This is that, if system A remains in equilibrium when isolated and placed in thermal contact first with system B and then with system C, the equilibrium of B and C will not be disturbed when they are placed in thermal contact with one another. For three systems in equilibrium at the same temperature, this is such a familiar circumstance that it was tacitly assumed in early expositions of heat theory, with the result that more rigorous contemporary accounts have to find room for it as the *Zeroth law of thermodynamics*.

Now suppose that a quantity of homogeneous gas (e.g. air) is trapped in a cylinder having diathermal walls and provided with a piston by which the volume V of the gas can be varied at will. If the pressure P applied to the piston is kept constant (e.g. at atmospheric value), the equilibrium state of the gas will be completely specified by its volume V. This system A can now be brought to a variety of states by being placed in a range of environments called heat baths or ovens. Given another system B in equilibrium, suppose the state of A is adjusted until the equilibrium of B is found not to be disturbed when it is placed in thermal contact with A. Then, we say that the *temperature* of B is measured by V. The instrument A is called a *gas thermometer*. It is important to note that it is being tacitly assumed that the temperature reading is unaffected by the manner in which thermal contact is established between A and B, i.e. that all parts of a system in equilibrium have the same temperature.

Clearly, if the temperature of any other system C is measured to have the same value V, then B and C are in equilibrium with the same system A and, by the zeroth law, must be in equilibrium with one another. We have therefore established the principle that, when two systems in equilibrium are found to have the same temperature, then they will remain in equilibrium when placed in contact with one another.

The precise nature of the standard system A used as a thermometer has no fundamental significance. If $(X_1, X_2, \ldots, X_n)$ is a complete set of parameters of state for a system S, by maintaining all but one constant in value, this one can be used as a measure of temperature. Then, if X, Y are the temperatures measured by two different thermometers S and T, a functional relationship $Y = f(X)$ must exist between them and the principle enunciated in the previous paragraph will apply with either measure.

The familiar mercury thermometer was originally employed to define several scales of temperature. The volume of the thread of mercury was measured against a scale of equal divisions marked on the evacuated container. In the case of the Celsius scale, the reading was taken to be 0 when the thermometer was in equilibrium with melting ice and to be 100 when it was in equilibrium with water boiling under standard atmospheric pressure. The Fahrenheit and Réaumer scales are of less importance, but will be found described in texts of elementary physics.

We have shown, therefore, that when the temperature scale associated with a thermometer has been established, every system in a state of equilibrium will possess a unique temperature θ. Thus, if $(X_1, X_2, \ldots, X_n)$ are the values of a complete set of parameters specifying the equilibrium state, then θ will be determined as a function of these variables, i.e.

$$\theta = \theta(X_1, X_2, \ldots, X_n). \tag{1.3.1}$$

Thus, θ and $(n-1)$ of the variables X_i constitute a complete set of state variables for, their values being given, the value of the remaining variable X_i is also fixed. In particular, for a homogeneous gas,

$$\theta = \theta(P, V). \tag{1.3.2}$$

Alternatively, this relationship can be expressed in either of the forms

$$P = P(V, \theta), \quad V = V(P, \theta), \tag{1.3.3}$$

showing that (V, θ) and (P, θ) are complete sets of state variables for the gas.

1.4   First law of thermodynamics. Internal energy

An early hypothesis regarding the nature of heat was that it was a 'subtle fluid' called *caloric* which flowed into a system when it was heated and flowed out of a system when it was cooled. This was refuted by James Joule (1818–89) who demonstrated that the temperature of an isolated system could be changed by the performance of work on the system alone. For example, he showed that the temperature of water can be raised by the rotation of a paddle wheel and, more precisely, that the rise in temperature on the Celsius scale is proportional to the work done by the wheel. The conclusion to be drawn from his experiments is that heat is a form of energy which, like kinetic energy, can be generated by the doing of work (we shall show, later, that statistical mechanics identifies the heat present in a system with the kinetic energy of its constituent molecules).

Joule showed further that the same change of state of a system was generated no matter how the given quantity of work was performed (e.g. the passage of an electric current through a resistor immersed in the water for the time needed to yield the same work done by the paddle, produced the same rise in temperature). From these, and other experiments, it was concluded that *the work needed to be performed on an isolated system to transfer it from one state of equilibrium to another depends only on the terminal states and is independent of the agency executing the work and the intermediate states through which the system passes.* This is the *First Law of Thermodynamics.*

It is a consequence of this first law that the *principle of conservation of energy* can be extended to apply to systems of the type we are considering. Thus, in the case of an isolated system, suppose work W is performed upon it to change its state from S_1 to S_2, the manner of the transformation being of no consequence. For energy to be conserved, it is necessary to suppose that the system possesses *internal energy U* and that this is increased by W. The first law fails to provide us with a means of determining U in any one state, but if this quantity is specified to be U_0 for some ground state or datum state S_0, its value in any other state S will be $U = U_0 + W$, where W is the work done in transferring the system between the states S_0 and S. From the more fundamental viewpoint of statistical mechanics, there is a natural choice for the state S_0, viz. the state in which all the system's molecules are in their quantum ground states and U_0 is the minimum energy eigenvalue for the system. This is not available to the classical theory, which must accept that the internal energy is arbitrary to the extent of an additive constant. However, once this has been fixed by specification of the internal energy in a datum state, U will be determinate for all other states and, hence, will be a function of the parameters of state, thus:

$$U = U(X_1, X_2, \ldots, X_n). \tag{1.4.1}$$

If the system under consideration is not isolated then, as we have remarked, its state will, in general, change when it is placed in thermal contact with other systems. In these circumstances, if work W is done on the system whilst it moves between two equilibrium states, the increment ΔU in its internal energy will not, necessarily, equal W; instead, we shall have an equation

$$\Delta U = W + Q. \tag{1.4.2}$$

The additional internal energy Q is called the *heat energy* supplied to the system. For example, the state of a beaker of water can be changed by operating a rotating paddle and, at the same time, placing its base in contact with the flame of a bunsen burner; the internal energy of the water will be increased by both the work done by the paddle and the heat supplied by the burner.

Next, suppose we have two systems in equilibrium states with internal energies U_1, U_2. Suppose these are placed in thermal contact in a container with adiabatic walls, so that the combined system is isolated from external influences. Initially, the total internal energy of the combined system is $U_1 + U_2$ and, if no work is done on either system, there can be no change in this energy when the systems interact. Hence, if $\Delta U_1, \Delta U_2$ are the energy changes for the systems when they have reached their new state of equilibrium, then

$$\Delta U_1 + \Delta U_2 = 0. \tag{1.4.3}$$

Since no work has been done on either system, equation (1.4.2) shows that $\Delta U_1 = Q_1$, $\Delta U_2 = Q_2$, Q_1, Q_2 being the heat energies supplied to the systems. Thus

$$Q_1 + Q_2 = 0; \tag{1.4.4}$$

i.e. the heat gained by one system must equal the heat lost by the other. In particular, when a system gains heat energy from its environment, the latter must lose an equal amount of heat. Evidently, the law of conservation of energy can be extended to systems of this type, provided heat energy is taken into account.

In such circumstances, where heat is transferred between systems in thermal contact, the system which loses heat energy is said to be *hotter* than the other system, or the system which gains energy is said to be *cooler*. As we have seen, if the two systems are at the same temperature, then no change will take place when they are placed in contact, i.e. there will be no transfer of heat. We now wish to show that, if transfer does take place, then we can always choose our temperature scale so that the temperature of the hotter system is invariably higher than that of the cooler system.

Suppose system A is hotter than system B and B is hotter than C. Then A is hotter than C, for if not, either A and C are equally hot or C is hotter than A. In the former case, make a small adjustment to the state of C so that it becomes hotter than A whilst still remaining cooler than B. Now bring A, B, C together into thermal contact, so that heat commences to flow from A to B, from B to C and from C to A. The rates of flow can be adjusted to be all the same by proper choice of the diathermal partitions separating the systems and then the combined system A + B + C will be in equilibrium. But as observed earlier, it is a basic supposition that all parts of a system in equilibrium are at the same temperature and must therefore be equally hot. We conclude that A must be hotter than C. It then follows, that systems can be arranged in order of hotness along an arbitrary scale and, since systems which are equally hot have equal temperatures and conversely, this scale can be used as a scale of temperature and it will be functionally related to any other scale associated with a thermometer. It will be assumed in future that such a hotness scale of temperature is being employed and that hotter systems have higher temperatures. It will be regarded as an observed fact that the Celsius scale satisfies this requirement.

1.5 Ideal gases

R. Boyle (1626–91) discovered the law which bears his name, viz. that, *at constant temperature* θ, *the volume* V *of a gas is inversely proportional to its pressure* P, or

$$PV = F(\theta). \tag{1.5.1}$$

This is very accurately obeyed by all gases or mixtures of gases resting in equilibrium states which are far from liquefaction.

This law was subsequently extended by L. Gay-Lussac (1778–1850), whose observations led to the conclusion that *the volume of a gas kept at constant pressure, is proportional to its absolute temperature*. The *absolute temperature* of a system is normally measured in kelvin (K) and is its temperature in degrees Celsius (°C) plus 273.16. Thus, if T is the absolute temperature of a gas,

$$V = T \times G(P). \tag{1.5.2}$$

An *ideal gas* is a hypothetical system which obeys both these laws perfectly. Its equation of state is accordingly

$$PV = CT, \tag{1.5.3}$$

where C is a constant for the gas sample under consideration.

At a given temperature and pressure, the volume of a given sample of gas is expected to be proportional to the number of molecules it contains. According to A. Avogadro (1776–1856), the constant in this proportionality is independent of the type of gas, i.e. *equal volumes of all gases at a given temperature and pressure contain equal numbers of molecules*. It follows that C (equation 1.5.3)) is proportional to the number of molecules N in the gas sample under consideration and hence that the equation of state for an ideal gas can be written

$$PV = kNT, \tag{1.5.4}$$

where k is a universal constant called *Boltzmann's constant* (its value in SI units is 1.38066×10^{-23} J K^{-1} (joules per kelvin)).

Suppose all the gas molecules are identical, i.e. the gas is not a mixture. Let w be their molecular weight. Then a *mole* of the gas is defined to have mass w grammes. Thus, a mole of any gas will have a specific number of molecules; this number is denoted by L and is called *Avogadro's number* ($L = 6.0221 \times 10^{23}$). A sample of gas comprising v moles will accordingly contain vL molecules and its equation of state (1.5.4) will be

$$PV = kvLT = vRT, \tag{1.5.5}$$

where $R = kL$ is termed the *gas constant* (8.3145 J mol^{-1} K^{-1}).

In the case of a mixture of v_1, v_2 moles of two different pure ideal gases which do not react chemically, each component contributes to the pressure as though it alone occupied the volume V (*Dalton's law*). Thus, if P_1, P_2 are the partial pressures, then

$$P_1 V = v_1 RT, \quad P_2 V = v_2 RT. \tag{1.5.6}$$

Addition shows that the pressure $P = P_1 + P_2$ of the mixture satisfies the equation of state (1.5.5) with $v = v_1 + v_2$. Clearly, this equation of state is valid for a mixture of any number of chemically inert gases, provided v is taken to be the sum of the masses of the constituents expressed in moles.

Next consider the internal energy of an ideal gas. An experiment first performed by Joule indicates that this is a function of the temperature alone. A container, with adiabatic walls, is divided into two compartments. One compartment is filled with gas and the other is evacuated. A valve in the dividing wall is now opened and the gas expands against zero pressure into the empty compartment. When equilibrium has been restored, the temperature of the gas is found to be unchanged from its erstwhile temperature prior to expansion. Since the gas expands against zero pressure, no work is performed on or by the gas. Further, since the container's walls are adiabatic, no heat energy is communicated to the gas. Hence, in equation (1.4.2), $W = Q = 0$ and, thus, $\Delta U = 0$. But (V, T) is a complete set of parameters of state for the gas (we use absolute temperature T from this point) and the internal energy U can accordingly be expressed as a function $U(V, T)$. In the Joule experiment, the initial and final volumes of the gas are different, but T is the same in both states. We conclude that U is a function of T alone.

Again, any actual gas fails to conform exactly to this law; if the thermometer employed is sufficiently sensitive, a slight fall in temperature will be detected. This is due to internal work being done in the expansion against the very small forces of attraction between the gas molecules—since the internal energy is conserved, the kinetic energy of the molecules' motion must be reduced to compensate, and this is felt as a fall in temperature (see section 2.4).

Suppose a gas is held at constant volume V and a steadily increasing quantity Q of heat is fed to it through the walls of its container so that the temperature T rises. We assume the influx of heat energy is controlled so that, at all instants, the gas remains virtually in an equilibrium state: i.e. the variation in state is quasi-static. Then $Q = Q(T)$ and, for any temperature T, the *heat capacity of the gas at constant volume* is defined to be C_V where

$$C_V = dQ/dT. \tag{1.5.7}$$

Since no work is performed on or by the gas, equation (1.4.2) indicates that

$$U = Q + \text{constant.} \tag{1.5.8}$$

Hence

$$C_V = (\partial U/\partial T)_V. \tag{1.5.9}$$

In the case of an ideal gas, U is independent of V and therefore

$$C_V = dU/dT. \tag{1.5.10}$$

The methods of statistical mechanics lead to the conclusion that, for an ideal gas, U is proportional to T and hence that C_V is constant. For one mole of a monatomic ideal gas, it will be shown (section 4.7) that $C_V = 3R/2$.

If the gas is held at constant pressure, the influx of heat will cause it to expand and work will be done against the external forces applied to the container. Suppose the container expands by a small volume dV, during which an element dS of its surface S suffers a small displacement dx. If $\mathbf{n}$ is the unit outward normal to dS, the volume traced out by this element is $(\mathbf{n} \cdot d\mathbf{x}) dS$; hence,

$$dV = \int_S (\mathbf{n} \cdot d\mathbf{x}) dS \tag{1.5.11}$$

The force applied to the element by the gas is $P\mathbf{n} \, dS$ and the work done by this force is therefore $P(\mathbf{n} \cdot d\mathbf{x}) dS$. The total work done by the gas is accordingly

$$dW = P \int_S (\mathbf{n} \cdot d\mathbf{x}) dS = P \, dV. \tag{1.5.12}$$

If we assume the expansion is quasi-static, the external forces applied to the container must balance the gas pressure and the work they do on the gas is therefore $-P\,dV$. Thus, if dQ is the heat supplied to the gas during the process, equation (1.4.2) gives

$$dU = dQ - P\,dV. \tag{1.5.13}$$

The *heat capacity at constant pressure* for the gas is denoted by C_P and is defined by

$$C_P = (dQ/dT)_P \tag{1.5.14}$$

Equation (1.5.13) then shows that

$$C_P = \{(dU + P\,dV)/dT\}_P = (\partial U/\partial T)_P + P(\partial V/\partial T)_P. \tag{1.5.15}$$

For an ideal gas, U depends on T alone and, therefore, $(\partial U/\partial T)_P = C_V$. Also, $V = vRT/P$, giving $P(\partial V/\partial T)_P = vR$. Hence

$$C_P = C_V + vR, \tag{1.5.16}$$

or, for one mole of gas, $C_P = C_V + R$. Evidently C_P, like C_V, is constant for an ideal gas.

Another quasi-static process which can usefully be studied to illustrate thermodynamic ideas is the adiabatic expansion of an ideal gas in a container whose walls are impermeable to heat. We can either suppose these walls to be elastic, so that the volume they contain can be varied as required, or that the volume can be altered by moving a piston in a cylinder. Consider a small change in the state of the gas in which the volume increases (slowly) from V to $V + dV$. Differentiating equation (1.5.5), the consequent changes in P and T must satisfy

$$V\,dP + P\,dV = vR\,dT. \tag{1.5.17}$$

But since no heat is supplied from an external source, equations (1.5.13), (1.5.10) show that

$$-P\,dV = dU = C_V\,dT \tag{1.5.18}$$

Elimination of dT between equations (1.5.17), (1.5.18) leads to the equation

$$dP/P = -\gamma\,dV/V, \tag{1.5.19}$$

where $\gamma = (C_V + vR)/C_V = C_P/C_V$ (equation (1.5.16)). Integration now yields the state equation

$$PV^\gamma = \text{constant} \tag{1.5.20}$$

governing the adiabatic expansion of an ideal gas by a quasi-static process.

We have here confined our calculations to quasi-static processes, since more general processes require us to follow a system through a succession of non-equilibrium states, to which the equations of equilibrium thermodynamics do not apply. However, it is sometimes useful to consider a dynamic process and the Joule experiment described above provides an example; evidently whilst the gas is flowing through the valve its state is not static and its expansion cannot be discussed using the equations of this section. Nevertheless, the first law is applicable to such a process, to relate the terminal equilibrium states. Further, it may be convenient, for the purposes of calculation, to imagine that the system is transferred between its terminal states by a quasi-static process; in such a hypothetical process, the intermediate states will bear little relationship to those in the actual process and, in particular, the manner in which heat is extracted or supplied to the system during the quasi-static process may be very different from that followed in the actual process (e.g. the actual process may be adiabatic and the hypothetical process not).

It is an important feature of an equilibrium state that it is *accessible*. By this we mean that the system can always, in principle, be prepared in the state by appropriate manipulation of the system's environment. This follows from the circumstance that such a state is prescribed by the values of a complete set of parameters of state and these physical quantities (pressure, volume, temperature, magnetization, etc.) are under the control of a human opera or. This implies that the system can be driven quasi-statically along any prescribed path through its state space and, in particular, the sense of the movement along a given path can be reversed. For this reason, a quasi-static process is often referred to as a *reversible process*.

On the other hand, a process which is not quasi-static is not under the control of any operator whose influence is restricted to the manipulation of a system's environment, for the motion of a system cannot be prescribed to follow a predetermined path by choice of the boundary conditions alone. Thus, once the valve separating the two compartments in the Joule experiment has been opened, the behaviour of the expanding gas is largely beyond external control and the succession of states through which the system moves is determined by the principles of fluid mechanics. In particular, the reverse process in which the gas flows back into one compartment can never be induced. Thus, non-quasi-static processes are generally *irreversible*.

A further example of an irreversible process is provided by Joule's experiment demonstrating that heat is a form of energy. The rotating paddle wheel does work on the water and transfers it from one equilibrium state to another at a higher temperature. However, the process is clearly not quasi-static and it is impossible to *lower* the temperature of the water by manipulation of the wheel, i.e. the process is irreversible.

1.6   Second law of thermodynamics. Absolute temperature

According to the first law, heat is a form of energy and may accordingly be created by performing work upon a system. Conversely, a system may itself do work and lose an equivalent amount of heat in the process; e.g. a hot gas may expand and drive a piston along a cylinder, thereby doing work and becoming cooler. A steam engine expands water vapour for this purpose and is constructed to work cyclically so that the process of expansion can be repeated indefinitely, resulting in a steady output of useful work. However, in such a case, although the engine's motion is cyclic, the complete thermodynamic system involved does not return to its original state at the end of each cycle—during each cycle, a quantity of steam is expanded, cooled and condensed and expelled into the atmosphere, and as the motion proceeds, a steadily increasing volume of condensed steam accumulates. No engine has ever been devised which, operating in a cycle, extracts heat from a reservoir and performs an equivalent amount of work, *the complete thermodynamic system (of which the engine is a part) returning to its initial state at the end of each cycle*. As an example, we know of no system of this type whose operation would have no other effect than to extract heat from the atmosphere and to perform an equivalent amount of work. The construction of such a system would permit the performance of work to continue until all sources of heat in the environment had been tapped to exhaustion and their temperatures reduced to absolute zero. Such a possibility is referred to as *perpetual motion of the second kind* and is denied by the *Second Law of Thermodynamics*. Perpetual motion of the first kind would occur if the engine were to perform work without the injection of an equivalent amount of heat and is, of course, forbidden by the first law.

Thus, Lord Kelvin's formulation of the second law is: *There exists no thermodynamic system which, being taken around a cycle of states, produces no effect except the extraction of heat from a reservoir and the performance of an equivalent amount of work.*

Consider any thermodynamic system H whose parameters of state are P (pressure), V (volume) and further quantities, the set of which will be denoted by X. Its temperature θ, on any convenient scale, will be expressible as a function $\theta(P, V, X)$ of these parameters and all possible states at a fixed temperature $\theta = \theta_0$ will be represented in the state space by a hypersurface whose equation is

$$\theta(P, V, X) = \theta_0. \qquad (1.6.1)$$

This hypersurface is called an *isotherm*.

Suppose H is prepared in the state (P_1, V_1, X_1) and is then placed in thermal contact with a heat reservoir at its own temperature θ_1. It is now permitted to expand quasi-statically to the state (P_2, V_2, X_2), remaining at the temperature θ_1, and extracts heat Q_1 from the reservoir; the path of the point in state space representing H's state will be a curve AB lying in the isotherm θ_1 (see Fig. 1.1). H is next isolated from the reservoir and further expanded adiabatically to the state (P_3, V_3, X_3), as represented by the curve BC in the state space. During this expansion, its temperature falls to θ_2. It is then placed in thermal contact with a second heat reservoir at temperature θ_2 and is compressed quasi-statically so that its state is changed to (P_4, V_4, X_4); the associated path in state space is the arc CD lying in the isotherm θ_2. During this compression, suppose heat Q_2 is transferred to the reservoir. H is again isolated and compressed adiabatically so that it is returned to the initial state (P_1, V_1, X_1) (path DA). It is assumed that the intermediate states B, C, D can be so chosen that the cycle closes. Such a cycle is called a *Carnot cycle* (S. Carnot, 1796–1832). Over the cycle, the net intake of heat is $Q_1 - Q_2$ and, by the first law, since the initial and final internal energies of the system must be the same, this must equal the work done by H on its environment.

Now suppose a second system K is transported around a Carnot cycle between two heat reservoirs at the same temperatures θ_1 and θ_2. Let Q'_1 be the heat

12

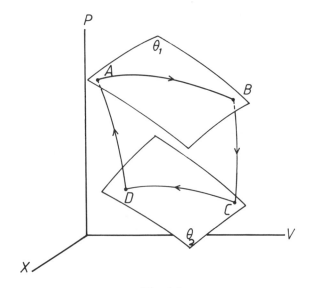

Fig. 1.1

extracted from the hotter reservoir and Q_2' be the heat delivered to the cooler reservoir. Suppose, if possible, $Q_1'/Q_2' > Q_1/Q_2$. Choose integers m and n such that $mQ_2 = nQ_2'$ (we can always suppose the vertices C, D of the path in state space are slightly adjusted as necessary to ensure Q_2, Q_2' are commensurable). Now set up a compound system L comprising m systems of the first type H and n of the second type K, all isolated from one another. Take each system H around its Carnot cycle in the opposite sense to that described earlier (recall that the cycle is reversible) and each system K around its Carnot cycle as described. Since we require the process to be quasi-static, the times during which the subsystems are in contact with the reservoirs can easily be synchronized; thus the system L completes a Carnot cycle. The total work done by the H-systems is $-m(Q_1 - Q_2)$ and by the K-systems in $n(Q_1' - Q_2')$ and hence the total work done by L is $n(Q_1' - Q_2') - m(Q_1 - Q_2) = nQ_1' - mQ_1$. This is positive since

$$nQ_1' = mQ_2 \cdot \frac{Q_1'}{Q_2'} > mQ_2 \cdot \frac{Q_1}{Q_2} = mQ_1. \tag{1.6.2}$$

Further, the heat transferred to the cooler reservoir by the H-systems is $-mQ_2$ and by the K-systems is nQ_2'; thus, the heat transferred by L to this reservoir is zero.

According to the hypothesis $Q_1'/Q_2' > Q_1/Q_2$, therefore, L is a system which can be operated cyclically to perform work by the simple extraction of heat from a single reservoir. This is in conflict with the second law and the hypothesis must accordingly be rejected. But the contrary hypothesis $Q_1/Q_2 > Q_1'/Q_2'$ can be shown to be untenable by transporting L around its Carnot cycle in the opposite sense. We have to conclude that $Q_1/Q_2 = Q_1'/Q_2'$ and hence that the ratio Q_1/Q_2

is independent of the type of system and can depend only on the temperatures θ_1, θ_2. Thus

$$Q_1/Q_2 = f(\theta_1, \theta_2), \tag{1.6.3}$$

where $f(\theta_1, \theta_2)$ is a universal function.

Let $\theta_1 > \theta_2 > \theta_3$ and suppose a system is taken around the Carnot cycle $ABCDA$ (Fig. 1.2) operating between the temperatures θ_1 and θ_2, extracting heat Q_1 from the hotter reservoir and surrendering heat Q_2 to the cooler reservoir. If the same system is carried around the Carnot cycle $DCEFD$ operating between temperatures θ_2 and θ_3, it will remove heat Q_2 from the hotter reservoir and deliver heat Q_3 to the cooler reservoir. If, finally, the system follows the Carnot cycle $ABEFA$ operating between the temperatures θ_1 and θ_3, heat Q_1 will be taken from the hotter reservoir and heat Q_3 will be expelled to the cooler. The following equations will now be valid:

$$Q_1/Q_2 = f(\theta_1, \theta_2), \quad Q_2/Q_3 = f(\theta_2, \theta_3), \quad Q_1/Q_3 = f(\theta_1, \theta_3). \tag{1.6.4}$$

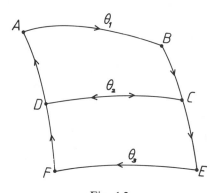

Fig. 1.2

We deduce that the function f must obey the multiplicative law

$$f(\theta_1, \theta_3) = f(\theta_1, \theta_2) f(\theta_2, \theta_3). \tag{1.6.5}$$

Giving θ_2 an arbitrary fixed value α, we deduce that $f(\theta_1, \theta_3) = g(\theta_1) \cdot h(\theta_3)$, where $g(\theta_1) = f(\theta_1, \alpha)$, $h(\theta_3) = f(\alpha, \theta_3)$. The identity (1.6.5) can accordingly be written in the form

$$g(\theta_1) \cdot h(\theta_3) = g(\theta_1) \cdot h(\theta_2) \cdot g(\theta_2) \cdot h(\theta_3). \tag{1.6.6}$$

It follows that $h(\theta_2) = 1/g(\theta_2)$ and, therefore, that

$$f(\theta_1, \theta_2) = g(\theta_1)/g(\theta_2). \tag{1.6.7}$$

Having determined the universal function $g(\theta)$, we use it to define the *absolute scale of temperature*. The transformation

$$T = g(\theta) \tag{1.6.8}$$

relates the temperature θ on the old scale to the temperature T on the absolute scale. Then, if temperatures θ_1, θ_2 on the old scale correspond to temperatures T_1, T_2 on the new scale, equation (1.6.3) can be written

$$Q_1/Q_2 = g(\theta_1)/g(\theta_2) = T_1/T_2. \tag{1.6.9}$$

It is evident from equation (1.6.7) that the function $g(\theta)$ is arbitrary to the extent of a constant multiplier. Thus the magnitude of the unit or degree in the absolute scale remains to be fixed. This we do by requiring that a rise of $100°$ in absolute temperature separates the temperatures of melting ice and boiling water, in the expectation that this will yield a unit which can be identified with the degree Celsius.

As a special case, suppose we transport an ideal gas, whose state equation is (1.5.5), around a Carnot cycle operating between the absolute temperatures T_1 and T_2 $(T_1 > T_2)$. During the isothermal expansion from state (P_1, V_1) to state (P_2, V_2), since the temperature T_1 is constant, the internal energy does not change. The heat extracted from the reservoir must accordingly equal the work done by the gas, i.e.

$$Q_1 = \int_{V_1}^{V_2} P\,dV = vRT_1 \ln(V_2/V_1). \tag{1.6.10}$$

Similarly, during the isothermal compression from state (P_3, V_3) to state (P_4, V_4), the heat delivered to the reservoir at temperature T_2 is the work done on the gas, viz.

$$Q_2 = -\int_{V_3}^{V_4} P\,dV = vRT_2 \ln(V_3/V_4). \tag{1.6.11}$$

Applying equation (1.5.5) to the isothermal phases, we derive the equations

$$P_1 V_1 = P_2 V_2, \quad P_3 V_3 = P_4 V_4. \tag{1.6.12}$$

Similarly, applying equation (1.5.20) to the adiabatic phases, we find

$$P_2 V_2^\gamma = P_3 V_3^\gamma, \quad P_4 V_4^\gamma = P_1 V_1^\gamma. \tag{1.6.13}$$

Multiplication of the equations (1.6.12) and then of the equations (1.6.13), yields

$$P_1 P_3 V_1 V_3 = P_2 P_4 V_2 V_4, \quad P_1 P_3 V_1^\gamma V_3^\gamma = P_2 P_4 V_2^\gamma V_4^\gamma. \tag{1.6.14}$$

Dividing these equations, we show that

$$V_1 V_3 = V_2 V_4. \tag{1.6.15}$$

It now follows that $V_2/V_1 = V_3/V_4$ and, hence, by equations (1.6.10), (1.6.11), that

$$Q_1/Q_2 = T_1/T_2. \tag{1.6.16}$$

This last equation is identical with equation (1.6.9) and we are therefore entitled to identify the absolute temperature defined in section 1.5 with that established in this section on the basis of the second law.

Finally, in this section, we give an alternative formulation of the second law due to R. Clausius (1822–88). This states that *no thermodynamic system can be constructed which, being taken around a cycle of states, produces no overall effect except the transfer of heat from a cooler to a hotter reservoir.*

For suppose such a system to be available and let Q be the heat transferred during each cycle from a reservoir at temperature T_2 to another at a higher temperature T_1. Couple this system to an engine which performs a Carnot cycle between the reservoirs in such a way that heat Q is extracted from the hotter reservoir and heat Q' is delivered to the cooler, work $(Q - Q')$ being performed in the process. It is assumed that the cycles of the two systems are synchronized. Clearly, at the end of a cycle of the dual system, the hotter reservoir has returned to its initial state. Thus, the resultant effect of the cycle is to extract heat $(Q - Q')$ from the cooler reservoir and to perform an equivalent amount of work. This is contrary to Kelvin's form of the second law and demonstrates that the Clausius statement of this law is perfectly equivalent.

1.7   Entropy. Clausius's inequality

In the last section, we studied the behaviour of a system as it was transported quasi-statically around a Carnot cycle. We shall now widen our consideration to permit the cycle to be of any type and also to be executed irreversibly if desired. The only condition on the cycle of states through which the system passes, therefore, is that the system should return precisely to its initial state after each round of the cycle.

As the system A is taken around its cycle, it will extract heat from or transfer heat to some external source, whose absolute temperature will be denoted by T (not necessarily constant). At any point in the cycle, let Q represent the total heat which has been extracted from the source (Q will be negative if a net quantity of heat has been delivered by A). We shall suppose the external source to be a system B of the type considered in the last section and that this executes a succession of Carnot cycles in the following manner: B is initially at a temperature T_0 and is then brought adiabatically to the temperature T. It is then placed in contact with A and compressed (or expanded) in such a way that it delivers heat dQ to A, whilst itself remaining at temperature T. During this process, A moves over an element of its cycle. B is next returned, adiabatically, to the temperature T_0, after which it is placed in contact with a reservoir at temperature T_0 and expands (or is compressed) isothermally to its initial state. Since B executes a Carnot cycle, equation (1.6.9) shows that the heat which must be extracted from the reservoir during this latter process is $T_0 \, dQ/T$. The whole process is imagined repeated for each element of A's cycle until this has been completed.

Let us now consider the combined system A + B. This is isolated, except for its contact with the reservoir and the total heat extracted from this reservoir over the whole cycle is $T_0 \int dQ/T$. But since the combined system is returned to its initial state at the end of the cycle, by the first law, this integral must equal the work done. We accordingly have a system operating in a cycle, which generates no other effect than to extract heat from a reservoir and to perform an equivalent amount of work. This is contrary to the second law, unless

$$\oint \frac{dQ}{T} \leqslant 0. \tag{1.7.1}$$

This is *Clausius's Inequality*. It applies to any cycle of a thermodynamic system, reversible or irreversible, where T is the temperature of the *heat source* (in the case of an irreversible cycle, it is possible for the temperatures of the source and system to differ).

In the special case when the system A is transported quasi-statically around a reversible cycle, the above argument can be repeated for the same cycle executed in the reverse sense. The heat extracted from the reservoir will then be $-T_0 \oint dQ/T$ and the second law requires that the inequality sign in (1.7.1) be reversed. We conclude that, for a reversible cycle,

$$\oint \frac{dQ}{T} = 0. \tag{1.7.2}$$

(Here, T is also the temperature of the system itself.)

This is a result of fundamental importance, since it leads to the definition of a new function of state called the *entropy*.

Let $X_i (i = 1, 2, \ldots, n)$ be a complete set of parameters of state for a thermodynamic system, interpreted as rectangular Cartesian coordinates of a point in state space. Suppose the system is moved quasi-statically from an arbitrarily chosen datum point D in its state space to a variable point X with coordinates X_i, extracting heat from its environment, as necessary, at a temperature T (variable). Then the integral

$$\int_D^X \frac{dQ}{T} \tag{1.7.3}$$

will be independent of the path from D to X. This follows, since any pair of paths form a closed circuit and, taken around such a circuit, the integral must vanish by equation (1.7.2); thus the contributions of the two parts of the circuit must cancel. We can now define a function of state $S(X_1, X_2, \ldots, X_n)$ by the equation

$$S = \int_D^X \frac{dQ}{T}. \tag{1.7.4}$$

S is the *entropy* in the state X and is arbitrary to the extent of an added constant (until D has been specified more precisely in section 5.5.)

It should here be noted that Q has been defined along a path but, being dependent upon the path, is not a function of state. The differential form of equation (1.7.4) is $dS = dQ/T$, indicating that $1/T$ is a factor (integrating factor) which transforms the differential dQ into an *exact differential* dS, which is expressible in terms of the parameters of state alone.

As an example, consider the case of an ideal gas. If, during a quasi-static process, its temperature rises by dT, equation (1.5.10) shows that its internal energy increases by $C_V \, dT$. This energy increase must equal the heat dQ supplied plus the work done on the gas; hence, by equation (1.5.13)

$$C_V \, dT = dQ - P \, dV. \qquad (1.7.5)$$

It now follows that the entropy is given by

$$S = \int_D^X \frac{dQ}{T} = C_V \int_{T_0}^T \frac{dT}{T} + \int_{V_0}^V \frac{P}{T} \, dV,$$

$$= C_V \ln T + \nu R \ln V + \text{constant}, \qquad (1.7.6)$$

having used the equation of state $PV = \nu RT$. Observe that the integration is independent of the path, as expected, and that S is expressed as a function of the state variables (V, T).

Since no heat is either added or subtracted from a gas during an adiabatic process, $dQ = 0$ for such a process and the entropy of the gas remains constant.

1.8 Illustrative problems

In this section, we shall study a number of problems whose solutions can be found by application of the principles already explained.

Problem 1

ν_0 moles of an ideal gas are contained in an insulated chamber at pressure P_0 and temperature T_0. The gas slowly escapes through a valve into an insulated cylinder provided with a frictionless piston to which an external pressure P_1 ($< P_0$) is applied. Initially the volume enclosed by the piston is zero. When the piston comes to rest, calculate the number of moles of gas left in the chamber and its temperature. Find, also, the temperature of the gas in the cylinder.

Solution

The expansion of the gas is adiabatic, but not quasi-static, since there will be a pressure gradient across the valve until the pressure in the chamber has been reduced to P_1 at the end of the process. However, the expansion of that portion of the gas which never leaves the chamber will be quasi-static and must be governed by equation (1.5.20). Thus, if V_0, V_1 are its volumes at the beginning and end of the process

$$P_0 V_0^\gamma = P_1 V_1^\gamma. \qquad (1.8.1)$$

Further, if T_0, T_1 are its terminal temperatures, the equation of state (1.5.5) shows that

$$P_0 V_0 = \nu_1 R T_0, \quad P_1 V_1 = \nu_1 R T_1, \qquad (1.8.2)$$

where v_1 is the number of moles of gas which remain in the chamber (volume V_1) at the end. Dividing equations (1.8.2) and using equation (1.8.1), we now find that

$$T_1/T_0 = (P_1 V_1)/(P_0 V_0) = (P_1/P_0)^{(\gamma - 1)/\gamma}, \qquad (1.8.3)$$

thus determining T_1.

Initially, v_0 moles of gas are present in the chamber, whose volume is V_1. Thus, the equation of state reads

$$P_0 V_1 = v_0 R T_0. \qquad (1.8.4)$$

Eliminating V_1 using the second of equations (1.8.2), we get

$$v_1/v_0 = (P_1 T_0)/(P_0 T_1) = (P_1/P_0)^{1/\gamma}, \qquad (1.8.5)$$

fixing v_1.

Finally, consider the irreversible process undergone by the whole gas. Let T be the final temperature of the gas in the cylinder and V its volume. The work performed by the gas on the piston is $P_1 V$ and this must be balanced by an equal reduction in its internal energy. If c_V is the heat capacity of the gas per mole, the reduction in internal energy of the v_1 moles of gas which never leave the chamber is $v_1 c_V (T_0 - T_1)$; similarly, the reduction for the $(v_0 - v_1)$ moles which escape to the cylinder is calculated to be $(v_0 - v_1)c_V(T_0 - T)$. We can now write down the equation of energy conservation, viz.

$$v_1 c_V (T_0 - T_1) + (v_0 - v_1)c_V (T_0 - T) = P_1 V. \qquad (1.8.6)$$

The equation of state for the gas in the cylinder at the end of the process is $P_1 V = (v_0 - v_1)RT$ and it follows that

$$v_0 T_0 - v_1 T_1 = \gamma (v_0 - v_1)T, \qquad (1.8.7)$$

where $\gamma = c_P/c_V = (c_V + R)/c_V$. Substituting for v_1 and T_1 from equations (1.8.3), (1.8.5), we are led to the result

$$T = \frac{T_0}{\gamma} \cdot \frac{1 - (P_1/P_0)}{1 - (P_1/P_0)^{1/\gamma}}. \qquad (1.8.8)$$

Problem 2

An insulated chamber fitted with a valve is exhausted. The valve is opened and air slowly admitted to the chamber from the atmosphere at temperature T. Calculate the temperature of the air in the chamber when the flow ceases.

Solution

Again, the process is not reversible, but we can apply the principle of energy conservation to the terminal states of the air which ultimately occupies the chamber.

Suppose the temperature of this air rises during the process from T to T'. If C_V is its heat capacity, treating it as an ideal gas its internal energy increases by $C_V(T' - T)$. This must equal the work done by the surrounding air. Initially, suppose the volume of the air which ultimately enters the chamber is V; then the work done by the surrounding air in forcing it through the valve is PV, where P is the atmospheric pressure. The equation of energy is accordingly

$$C_V(T' - T) = PV. \tag{1.8.9}$$

The equation of state of the air before it enters the chamber is $PV = vRT$. It follows that

$$T' = (1 + vR/C_V)T = \gamma T. \tag{1.8.10}$$

Problem 3

A steel bar, whose heat capacity is h, is heated to temperature t and then plunged into an oil bath whose heat capacity is H and temperature is T. Assuming no loss of heat from the system 'bar + bath', calculate the increase in entropy.

Solution

If the process were quasi-static, since no heat is extracted or supplied to the system $(dQ = 0)$, the entropy would be conserved. However, the process is clearly irreversible and the entropy increases.

We first calculate the equilibrium temperature T' of the system. Equating the heat lost by the steel to the heat gained by the oil, we obtain the equation

$$(t - T')h = (T' - T)H. \tag{1.8.11}$$

(We have assumed h, H to be constants.) Thus,

$$T' = (TH + th)/(H + h). \tag{1.8.12}$$

Now consider a hypothetical quasi-static process in which the temperature of the oil is slowly raised from T to T'. During this process, as the temperature rises from θ to $\theta + d\theta$, the heat supplied must be $dQ = Hd\theta$ and the entropy increase is $dQ/\theta = Hd\theta/\theta$. The net entropy increase for the whole process is therefore

$$H \int_T^{T'} d\theta/\theta = H \ln(T'/T). \tag{1.8.13}$$

Since the entropy is a function of state, the manner in which a system is transferred between a pair of equilibrium states is irrelevant to the matter of entropy increase and we can therefore accept this last result as the actual entropy increase for the oil. Similarly, the entropy increase for the steel is found to be

$$h \int_t^{T'} d\theta/\theta = -h \ln(t/T'). \tag{1.8.14}$$

Thus, the net increase in entropy is given by

$$\Delta S = (H + h)\ln T' - H\ln T - h\ln t, \tag{1.8.15}$$

which can be proved to be always positive.

If, as is generally the case, the oil vat is very large, then h/H will be small and we can approximate thus:

$$\Delta S = (H + h)\left[\ln\{T(1 + th/TH)(1 + h/H)^{-1}\}\right] - H\ln T - h\ln t,$$
$$= (H + h)\left[\ln T + th/TH - h/H + 0\,(h^2/H^2)\right] - H\ln T - h\ln t,$$
$$= h[t/T - 1 - \ln(t/T)], \tag{1.8.16}$$

approximately.

Problem 4

One mole of each of two different ideal gases is held at a temperature T in a pair of identical chambers. The chambers are connected by a pipe and the gases diffuse into one another without reacting chemically. Calculate the entropy increase.

Solution

This process, also, is irreversible, since the diffusion process is a dynamic one and the intermediate states are not equilibrium states. However, the terminal states for the process are equilibrium states and we can calculate the entropy increase by supposing each gas to be transported to its final state quasi-statically (or, indeed, in any convenient manner).

Clearly the temperature of the mixture is the same as the initial temperature of the gases, viz. T. We consider an isothermal quasi-static process, therefore, in which the partial pressure P of a gas and its volume V are related by the equation of state $PV = RT$. During the process, the internal energy of the gas remains constant and $dQ = P\,dV$. Thus, its entropy increase is given by

$$\int_v^{2v} P\,dV/T = R\int_v^{2v} dV/V = R\ln 2, \tag{1.8.17}$$

v being the volume of a chamber. The entropy increase for the two gases is accordingly $2R\ln 2$.

In the special case where the two gases are identical, their final state is evidently thermodynamically indistinguishable from their initial state and there can be no change in entropy. The failure of the above argument to yield the correct result in these circumstances cannot be satisfactorily explained by appeal to classical ideas (we are confronted by the *Gibbs paradox*). According to quantum theory, however, immediately the two gases are connected, all their molecules become indistinguishable and it is meaningless to speak of the one gas diffusing into the other. This situation does not arise when the gases are chemically distinct and the diffusion process is then a reality and an acceptable feature of the argument.

Problem 5

Prove that the internal energy of an ideal gas is a function of its temperature alone.

Solution

This result has already been shown (section 1.5) to be a consequence of the first law and Joule's observation that, when a gas is permitted to expand into an evacuated enclosure, its temperature is unchanged. We now prove that this result is also derivable from the first two laws alone.

For any fluid, suppose its temperature increases by dT and its volume by dV. Then, since U can be expressed as a function of the complete set of state variables (T, V), the increase in the internal energy is given by

$$dU = (\partial U/\partial T)_V \, dT + (\partial U/\partial V)_T \, dV. \qquad (1.8.18)$$

But the first and second laws show that

$$dU = T \, dS - P \, dV \qquad (1.8.19)$$

and we conclude that

$$T \, dS = (\partial U/\partial T)_V \, dT + (P + (\partial U/\partial V)_T) \, dV. \qquad (1.8.20)$$

It now follows that

$$\left(\frac{\partial S}{\partial T}\right)_V = \frac{1}{T}\left(\frac{\partial U}{\partial T}\right)_V, \quad \left(\frac{\partial S}{\partial V}\right)_T = \frac{1}{T}\left\{P + \left(\frac{\partial U}{\partial V}\right)_T\right\} \qquad (1.8.21)$$

and, hence, by calculating $\partial^2 S/(\partial T \partial V)$ from both these results, we require that

$$\frac{\partial}{\partial V}\left(\frac{1}{T}\frac{\partial U}{\partial T}\right) = \frac{\partial}{\partial T}\left\{\frac{1}{T}\left(P + \frac{\partial U}{\partial V}\right)\right\}. \qquad (1.8.22)$$

This leads at once to the condition

$$\frac{\partial U}{\partial V} = T\frac{\partial P}{\partial T} - P. \qquad (1.8.23)$$

This result is applicable to any fluid, but if the fluid is an ideal gas whose equation of state is $PV = vRT$, then it reduces to $\partial U/\partial V = 0$, implying, of course, that U is a function of T alone. The methods of statistical mechanics (sections 3.7 and 7.2) show that $U = C_V T$, where C_V is a constant.

1.9 Law of increasing entropy

It has been shown (see (1.7.1)) that, if a thermodynamic system undergoes a cyclic process (reversible or irreversible), during which heat dQ is supplied from an external source at absolute temperature T, then for any complete cycle, Clausius's inequality is valid, viz.

$$\oint dQ/T \leqslant 0. \tag{1.9.1}$$

If the process is reversible, then the equality sign applies.

Now suppose that a process transfers a system from an equilibrium state A to another B. Let S_A and S_B be the system's entropies in the two states. If this process is followed by a quasi-static process which returns the system to state A, the two processes together will constitute a cyclic process for which the Clausius inequality is valid. But the contribution to the left-hand member of the inequality (1.9.1) of the quasi-static process will be $S_A - S_B$, by the definition of entropy. Hence, the inequality can be written

$$\int_A^B dQ/T + S_A - S_B \leqslant 0, \tag{1.9.2}$$

where the integral is to be calculated for the original process.
Rearranging the inequality (1.9.2), we find

$$\Delta S = S_B - S_A \geqslant \int_A^B dQ/T, \tag{1.9.3}$$

where ΔS is the entropy increase due to the original process.

In particular, if the system is thermally isolated from its environment, then $dQ = 0$ and the inequality (1.9.3) reduces to

$$\Delta S \geqslant 0. \tag{1.9.4}$$

Thus, *when a system executes any process which transfers it adiabatically from one equilibrium state to another, its entropy cannot decrease.* If the process is quasi-static, the entropy remains constant.

Problems 3 and 4, whose solutions have been given in the previous section, provide examples of the law of increasing entropy for an isolated system.

Thus far, we have only defined the entropy for systems which have attained thermal equilibrium. However, it is easy to extend the concept to systems which comprise a number of subsystems separated from one another by adiabatic walls and which are in thermal equilibrium individually at different temperatures (and pressures). Since the temperature of the composite system is not uniform, it is not itself, strictly, in thermal equilibrium and, because no real partition can be perfectly impervious to heat, in practice the temperatures of the subsystems will slowly tend to equalize. Nevertheless, it is often useful to study the ideal case and we then define the entropy of the combined system as being the sum of the entropies of the subsystems. In these circumstances, if the adiabatic walls are replaced by flexible diathermal membranes so that the systems can interact thermally and mechanically, provided the composite system is thermally isolated, its entropy will increase.

For, consider the case of a pair of subsystems A and B, where A is initially hotter than B. The process for the composite system which takes place when A and B interact is irreversible, since the combined system is never in equilibrium except at the end. However, each subsystem moves from one equilibrium state to another and, since only the terminal states of the composite system are in question, it is permissible to replace the actual process by one in which the subsystems move reversibly between their terminal states. In such a hypothetical process, the dividing wall will move and conduct heat very slowly, so that the states of A and B evolve quasi-statically (the process remains irreversible for the combined system). Suppose at some stage during this process A is at temperature T_A and loses heat dQ to B, which is at temperature T_B ($T_A > T_B$). Then A loses entropy dQ/T_A and B gains entropy dQ/T_B. Thus the composite system shows a net gain in entropy of $dQ(1/T_B - 1/T_A)$. Integrating this gain over the whole process, it is plain that the entropy of the composite system must increase. The extension of the entropy principle to isolated systems comprising more than two subsystems can be performed similarly.

Exercises 1

1. For any fluid, the isothermal compressibility k_T and the expansion coefficient β are given by

$$k_T = -\frac{1}{V}\left(\frac{\partial V}{\partial P}\right)_T, \quad \beta = \frac{1}{V}\left(\frac{\partial V}{\partial T}\right)_P.$$

If $k_T = 3(V-a)/(4PV)$, $\beta = (V-a)/(TV)$, show that the equation of state is $P^{3/4}(V-a) = AT$, where A is constant.

2. For a fluid, the isotherm through the critical point has zero slope and a point of inflexion. If the equation of state for a gas is Berthelot's equation, viz.

$$(P + a/TV^2)(V - b) = vRT$$

show that, at the critical point, $V_c = 3b$, $T_c = (8a/27vRb)^{1/2}$, $P_c = \frac{1}{12b}\left(\frac{2avR}{3b}\right)^{1/2}$.

3. An ideal gas is initially in the state (P_0, V_0). It is expanded adiabatically to the state (P_1, V_1). Show that the work done by the gas is

$$\frac{P_0 V_0}{\gamma - 1}\left[1 - \left(\frac{V_0}{V_1}\right)^{\gamma - 1}\right].$$

If it is first expanded isothermally to the volume V_1 and its pressure then brought to P_1 by a change in temperature, show that the work done is $P_0 V_0 \ln(V_1/V_0)$. If, instead, the gas is first cooled at constant volume V_0 to the pressure P_1 and then expanded by heating at constant pressure to the volume V_1, show that the work done is

$$P_0 V_0\left[\left(\frac{V_0}{V_1}\right)^{\gamma - 1} - \left(\frac{V_0}{V_1}\right)^{\gamma}\right].$$

4. A mole of an ideal gas is expanded from a volume V_0 to a volume V_1 by a process described by the equation $P = \alpha/V^2$. Show that the work done by the gas in the process is $\alpha(V_1 - V_0)/(V_0 V_1)$ and that the heat extracted from the gas is

$$\alpha \left(\frac{C_V}{R} - 1 \right) \left(\frac{1}{V_0} - \frac{1}{V_1} \right).$$

5. A litre of water is contained in a cylinder fitted with a piston. Calculate the work done when the pressure is increased quasi-statically and isothermally from 1 atmosphere to 200 atmospheres, it being assumed that the compressibility k_T (see Ex. 1) of water has constant value $4 \times 10^{-10} \, \text{m}^2/\text{N}$. (Assume 1 atm $= 101325 \, \text{N} \, \text{m}^{-2}$.) (Ans. 82.1 J.)

6. Show that, for an adiabatic process executed on a fluid,

$$C_P (\partial T/\partial P)_{\text{ad}} = -(\partial U/\partial P)_T - P (\partial V/\partial P)_T.$$

7. N ideal gases at the same temperature and pressure occupy separate containers whose volumes are $V_1, V_2, \ldots, V_N$. The containers are connected so that the gases diffuse into one another without any chemical action taking place. Show that the entropy increase is

$$-vR \sum_{i=1}^{N} x_i \ln x_i,$$

where $x_i = V_i/(V_1 + V_2 + \ldots + V_N)$ and v is the total number of moles in the mixture.

8. The Van der Waals' equation of state for a gas is

$$(P + a/V^2)(V - b) = vRT.$$

Employing the condition (1.8.23), prove that

$$U = f(T) - a/V,$$

where $f(T)$ is a function of T alone. Deduce that C_V depends on T alone.

9. Replacing the equation of state in the previous problem by Berthelot's equation (see Ex. 2), show that

$$C_V = g(T) + 2a/T^2 V.$$

10. 1 kilogram of nickel is heated so that its temperature rises from $20 \, ^\circ\text{C}$ to $300 \, ^\circ\text{C}$. Assuming its specific heat at constant pressure c_P has the constant value $443 \, \text{J} \, \text{kg}^{-1} \, \text{K}^{-1}$, calculate its entropy increase. (Hint: Bring the metal from its initial to its final state by a reversible process at constant pressure.) (Ans. $296 \, \text{J} \, \text{K}^{-1}$.)

11. v moles of an ideal gas are contained in an insulated cylinder of volume V_1 and the gas is permitted to expand through a valve into an evacuated and insulated chamber having volume V_2. Explain why the temperature of the gas is the same at the end of the process as at the beginning and show that the increase in entropy is $vR \ln(1 + V_2/V_1)$. (Hint: Replace the actual process by a quasi-static isothermal process.)

12. The internal energy of a certain Van der Waals gas is given by $U = C_V T - a/V$. Initially it is at a temperature T_1 and occupies a volume V_1. It is allowed to expand adiabatically into a vacuum to a final volume V_2. Show that its temperature falls by

$$\frac{a}{C_V}\left(\frac{1}{V_1} - \frac{1}{V_2}\right)$$

and give the physical reason for the fall.

CHAPTER 2

Applications of classical theory

2.1 Two-parameter systems

For many thermodynamic systems, a complete set of state variables comprises only two quantities. A homogeneous fluid provides the most familiar example, since any pair of the variables (P, V, T) is a complete set. However, many other systems of this type are encountered in thermophysics and a number of examples will first be given, before applying to them the general theory developed in the previous chapter.

A wire stretched by equal and opposite forces P applied at its ends can be cited. In addition to P, its length L and absolute temperature T are parameters of state and these three variables are related by an equation of state which, for many materials, may be approximated as $P = \mu T(L - L_0)$, where μ is constant. Any two of these parameters yields a complete set. If the state changes infinitesimally and quasi-statically so that L increases by dL, the work done on the wire is $P\,dL$ and the associated energy equation is

$$dU = T\,dS + P\,dL. \tag{2.1.1}$$

More precisely, the extension of the wire will be accompanied by a contraction in its cross-sectional area A (related to the extension by the elastic constant called *Poisson's ratio*) and the atmospheric pressure Π acting upon the wire's surface will accordingly do further work and this will contribute a term to the right-hand member of equation (2.1.1). Thus, Π and A should be included amongst the parameters of state and the system would then cease to be a two-parameter one. However, this correction to equation (2.1.1) is usually accepted to be negligible.

A soap film spread across a closed wire affords a second example. Convenient state parameters are the temperature T, the surface tension σ and the area A of the film. If the state of the film is changed by manipulation of the wire, the work done on the film for a reversible change is $2\sigma\,dA$ (NB the film has two surfaces) and the energy equation takes the form

$$dU = T\,dS + 2\sigma\,dA. \tag{2.1.2}$$

Again, the parameters T, σ, A are related by an equation of state.

A third illustration of a two-parameter system is a block of magnetic material placed in an externally controlled magnetic field. Its magnetization **I** (i.e. moment per unit volume) at a point can be controlled by variation of the magnetic intensity

H at the point ($I = \chi H$, where χ is the material's *magnetic susceptibility*). H will be the resultant of the applied external field and the field due to the induced magnetization. Of these two components, only the first can be controlled directly. For paramagnetic and diamagnetic substances at normal temperatures, the contribution of the induced magnetism is small and is usually neglected in computing H; but for paramagnetic substances obeying *Curie's law* ($\chi = C/T$), at temperatures close to absolute zero such neglect is not justified. In the case of ferromagnetic materials, the magnitude of the induced field is always comparable (and opposed) to the applied field and must be taken into account.

It follows (see Appendix A) from Maxwell's equations of the electromagnetic field that, if the field quantities H, B change by increments dH, dB, then the total electromagnetic energy density increases by $H \cdot dB$. But $B = \mu_0(H + I)$, so that the increment in the energy density is $\mu_0 H \cdot dH + \mu_0 H \cdot dI$. The first term in this expression gives the work done raising the energy of the field and the second term represents the work done creating additional magnetization dI. We shall ignore the field energy, since this makes no contribution to the energy of the block. The potential energy of the created magnetization in the field H is $d(-\mu_0 H \cdot I)$ and this will be included to give for the total increment in the energy density

$$dW = \mu_0 H \cdot dI - \mu_0 d(H \cdot I) = -\mu_0 I \cdot dH. \tag{2.1.3}$$

In particular, if we assume H and I are always parallel and constant in magnitude and direction over the magnet, then the increment in the block's magnetic energy is $-\mu_0 M \, dH$, where $M = VI =$ overall magnetic moment (V is the volume of the block). The energy equation for the block is then

$$dU = T \, dS - \mu_0 M \, dH \tag{2.1.4}$$

and we have a system with state variables T, H, M. The magnetic susceptibility $\chi = \chi(H, T)$ and, therefore, $M = M(H, T)$; thus (T, H) is a complete set of parameters of state and the system is a two-parameter one.

It is relevant at this stage to consider the phenomenon of *hysteresis*. This occurs when the physical characteristics of a system are dependent upon its past history. For example, the wire introduced earlier may be stretched beyond its elastic limit, in which case its elastic properties are changed irrevocably. In these circumstances, a quasi-static return to an earlier state will have become impossible, since we are now effectively operating upon a new system. Thus, even if the stretching was carried out quasi-statically, it would not be reversible. Similarly, if we first magnetize a piece of steel and then proceed to demagnetize it, even if the process is performed quasi-statically, the magnetization and demagnetization curves traced out in the $H-M$ plane will not be identical, i.e. again, the process is not reversible. Systems which are subject to hysteresis are very difficult to treat thermodynamically and will henceforth be excluded from our consideration.

In the general case where a complete set of parameters comprises more than two variables, we assume that a complete set $(X_1, X_2, \ldots, X_n)$ can be identified such that, when the state changes quasi-statically with X_i increasing by

$\mathrm{d}X_i (i = 1, 2, \ldots, n)$, the work done on the system can be expressed in the form

$$\mathrm{d}W = \sum_{i=1}^{n} F_i \mathrm{d}X_i, \qquad (2.1.5)$$

where the coefficients F_i can be expressed as functions of the X_i and are called the *generalized components of force* acting upon the system. An energy equation can then be written down and a mathematical analysis performed in a manner which is similar, but necessarily more complex, than the one we shall give for a two-parameter system in the sections which follow.

2.2 Enthalpy, free energy and Gibbs function

Until notice is otherwise given, the two-parameter system under analysis will be presumed to be a fluid for which the following state variables have now been defined: P, V, T, S, U. In general, the values of any two members of this set can be prescribed to fix the thermodynamical equilibrium state of the fluid; these values being given, the values of the remaining three variables are determinate—thus it may be assumed that any three of these state variables are functions of the other two. In special cases, there may be exceptions to this rule; e.g. in the case of an ideal gas, U is a function of T alone and, as a consequence, the set (U, T) is not complete.

The energy equation for a fluid, viz.

$$\mathrm{d}U = T\mathrm{d}S - P\mathrm{d}V, \qquad (2.2.1)$$

indicates that, if we regard (S, V) as a complete set of independent variables, then

$$\left(\frac{\partial U}{\partial S}\right)_V = T, \quad \left(\frac{\partial U}{\partial V}\right)_S = -P. \qquad (2.2.2)$$

The mixed second derivative $\partial^2 U / \partial S \, \partial V$ can now be calculated from both these results; since these must be identical, it is necessary that

$$\left(\frac{\partial T}{\partial V}\right)_S = -\left(\frac{\partial P}{\partial S}\right)_V. \qquad (2.2.3)$$

This is the first of *Maxwell's thermodynamical relations*.

Other relationships can be established in a similar manner by first defining three new variables of state, viz. the *enthalpy H*, the *Helmholtz free energy F*, and the *Gibbs function G*, by means of the equations

$$H = U + PV, \qquad (2.2.4)$$

$$F = U - TS, \qquad (2.2.5)$$

$$G = U + PV - TS. \qquad (2.2.6)$$

Clearly,

$$\mathrm{d}H = \mathrm{d}U + P\mathrm{d}V + V\mathrm{d}P = T\mathrm{d}S + V\mathrm{d}P. \qquad (2.2.7)$$

Thus, if (S, P) are chosen as the complete set of independent variables, then

$$\left(\frac{\partial H}{\partial S}\right)_P = T, \quad \left(\frac{\partial H}{\partial P}\right)_S = V, \tag{2.2.8}$$

Calculation of $\partial^2 H/\partial S\,\partial P$ from both these equations, now yields the second Maxwell relationship

$$\left(\frac{\partial T}{\partial P}\right)_S = \left(\frac{\partial V}{\partial S}\right)_P. \tag{2.2.9}$$

Similarly, from equations (2.2.5), (2.2.6), we find

$$dF = -S\,dT - P\,dV, \tag{2.2.10}$$
$$dG = -S\,dT + V\,dP, \tag{2.2.11}$$

from which we deduce the remaining two Maxwell relations, viz.

$$\left(\frac{\partial S}{\partial V}\right)_T = \left(\frac{\partial P}{\partial T}\right)_V, \tag{2.2.12}$$

$$\left(\frac{\partial S}{\partial P}\right)_T = -\left(\frac{\partial V}{\partial T}\right)_P. \tag{2.2.13}$$

(Note: As an aid to memory, observe that T, V, P, S occur in all four relations, and that formal cross-multiplication yields the dimensionally homogeneous products PV and TS; the signs are obvious if the physical significance of the derivatives involved is considered.)

Further (more complex) relations of this type can be established by choosing other pairs of state variables as the complete set (see e.g. Exercises 1–4 at the end of this chapter). Since all these identities are consequences of the same fundamental energy equation, they are not independent and may be derived from one another by application of standard techniques from the calculus of partial derivatives.

If U is given as a function of S and V, we can write down the equation

$$dU = (\partial U/\partial S)_V\,dS + (\partial U/\partial V)_S\,dV \tag{2.2.14}$$

and this must be identifiable with the energy equation (2.2.1). Thus equations (2.2.2) follow and T and P will be determined as functions of S and V. Eliminating S, we then obtain the equation of state and all thermodynamical properties of the fluid are calculable. Similarly, if $H = H(S, P)$ is known, equation (2.2.7) follows and, again, the properties of the fluid are specified.

$F = F(T, V)$ or $G = G(T, P)$ also serve (*via* equations (2.2.10) and (2.2.11)) to specify the fluid's thermodynamic behaviour completely. Noting that (T, V) are the most convenient fundamental set of parameters for the purposes of statistical mechanics (see Chapter 3), the free energy F becomes a most important means of specifying a fluid's properties. From equation (2.2.10), it follows that

$$S = -(\partial F/\partial T)_V, \quad P = -(\partial F/\partial V)_T, \tag{2.2.15}$$

and then, from equations (2.2.5), (2.2.6), we get

$$U = F - T(\partial F/\partial T)_V = -T^2 \left(\frac{\partial}{\partial T} \frac{F}{T}\right)_V, \qquad (2.2.16)$$

$$G = F + PV = F - V(\partial F/\partial V)_T = -V^2 \left(\frac{\partial}{\partial V} \frac{F}{V}\right)_T. \qquad (2.2.17)$$

H now follows from equations (2.2.4), (2.2.15) and (2.2.16).

Finally, in this section, it will be helpful to establish the physical *bona fides* of the new state variables H, F, G.

Suppose the fluid is transported quasi-statically and at constant pressure between two states. For such a change

$$\Delta H = \Delta U + P\,\Delta V = Q, \qquad (2.2.18)$$

where Q is the heat supplied. Thus, for a reversible process at constant pressure (e.g. atmospheric), the heat absorbed by the system equals the increase in the enthalpy. Thus the heat capacity at constant pressure is given by

$$C_P = (\partial H/\partial T)_P. \qquad (2.2.19)$$

For a given quasi-static process at constant temperature,

$$\Delta F = \Delta U - T\,\Delta S = \Delta U - Q = W, \qquad (2.2.20)$$

where W is the work done on the fluid. Hence, for such a process, the decrease in the free energy equals the work done by the fluid.

If the quasi-static process takes place with both temperature and pressure constant, the Gibbs function becomes physically significant. A two-parameter system cannot vary its state if both T and P are held constant, but suppose a complete set of state variables for the fluid (which may, perhaps, be magnetic) includes quantities X_i and the energy equation takes the form

$$dU = T\,dS - P\,dV + \sum F_i\,dX_i. \qquad (2.2.21)$$

Then a reversible process in which T and P are kept constant and the generalized components of force F_i do work W, becomes possible and, for such a process

$$\Delta G = \Delta U + P\,\Delta V - T\,\Delta S = W. \qquad (2.2.22)$$

That is, the increment in the Gibbs function measures the work done on the system by the forces F_i.

2.3 Experimental characteristics for a fluid

It was remarked in the previous section that if any one of the state variables U, H, F, G is known as a function of an appropriate pair of state variables T, P, V, S, then all thermodynamic properties of the fluid are calculable. However, experiment does not make such information directly available and it is necessary to consider what are the minimal observable data needed to permit a complete derivation of these properties.

Suppose we start by assuming that a determination has been made of the equation of state, so that any one of the variables (P, V, T) can be regarded as a known function of the other two. In particular, suppose $P = P(T, V)$ is known, so that the right-hand member of the Maxwell relation (2.2.12) is calculable. Then, integrating this equation with respect to V keeping T constant, we obtain

$$S(T, V) - S(T, V_0) = \int_{V_0}^{V} (\partial P / \partial T)_V \, dV, \tag{2.3.1}$$

where V_0 is an arbitrary initial value for the volume.

If the fluid's volume is maintained constant whilst its temperature and entropy increase quasi-statically by dT and dS respectively, then the heat which has been communicated to it is $T\,dS$ and its heat capacity at constant volume is given by

$$C_V = T(\partial S / \partial T)_V. \tag{2.3.2}$$

Hence, if C_V is known for one value of V as a function of T, this equation can be integrated with respect to T, with $V = V_0$, to yield

$$S(T, V_0) - S(T_0, V_0) = \int_{T_0}^{T} (C_V / T) \, dT. \tag{2.3.3}$$

Adding equations (2.3.1) and (2.3.3), we find

$$S(T, V) = S(T_0, V_0) + \int_{T_0}^{T} (C_V / T) \, dT + \int_{V_0}^{V} (\partial P / \partial T)_V \, dV, \tag{2.3.4}$$

showing that $S(T, V)$ is calculable, given its value in some datum state (T_0, V_0) (which can, at this stage, be chosen arbitrarily).

Knowing $S(T, V)$ and $P(T, V)$, equations (2.2.15) permit the calculation of $F = F(T, V)$ (except for an additive constant having no physical significance). All the properties of the fluid then become determinate.

Alternatively, we can first integrate equation (2.2.13) with respect to P at constant temperature to derive

$$S(T, P) - S(T, P_0) = - \int_{P_0}^{P} (\partial V / \partial T)_P \, dP. \tag{2.3.5}$$

For the heat capacity at constant pressure, equation (2.3.2) must be replaced by

$$C_P = T(\partial S / \partial T)_P. \tag{2.3.6}$$

Thus, integrating with respect to T keeping P constant at the value P_0, we get

$$S(T, P_0) - S(T_0, P_0) = \int_{T_0}^{T} (C_P / T) \, dT. \tag{2.3.7}$$

Addition of equations (2.3.5), (2.3.7) gives the function $S(T, P)$ in the form

$$S(T, P) = S(T_0 \ P_0) + \int_{T_0}^{T} (C_p/T)\mathrm{d}T - \int_{P_0}^{P} (\partial V/\partial T)_P \, \mathrm{d}P. \qquad (2.3.8)$$

From equation (2.2.11), we deduce that

$$(\partial G/\partial T)_P = -S, \quad (\partial G/\partial P)_T = V. \qquad (2.3.9)$$

Hence, knowing $S(T, P)$ and $V(T, P)$, $G(T, P)$ can be found and the system's properties are then calculable.

To summarize, given (i) a fluid's equation of state and (ii) either C_V as a function of T for a single value of V, or C_P as a function of T for a single value of P, then the fluid's properties are all derivable.

An example is provided by a gas whose equation of state is the Van der Waals equation

$$(P + a/V^2)(V - b) = \nu RT. \qquad (2.3.10)$$

This gives

$$(\partial P/\partial T)_V = \nu R/(V - b). \qquad (2.3.11)$$

If we assume C_V is constant for the volume V_0, then equation (2.3.4) shows that

$$S = C_V \ln T + \nu R \ln(V - b) + \text{constant}. \qquad (2.3.12)$$

Integrating equations (2.2.15), we now calculate that

$$F = C_V T(1 - \ln T) - \nu RT \ln(V - b) - \frac{a}{V} + \text{constant} \qquad (2.3.13)$$

and equation (2.2.5) then shows that the internal energy is given by

$$U = F + TS = C_V T - a/V + \text{constant}. \qquad (2.3.14)$$

2.4 Characteristics of a fluid

In this section, we shall introduce a number of important coefficients which measure various characteristics of a fluid and obtain some useful formulae from which they may be calculated when we are given the equation of state.

Having shown in the previous section that, if C_V is known on the line $V = V_0$ in the VT-plane, the properties of the fluid can be found from the equation of state, it is to be expected that C_V itself can be computed for all values of V by integration from V_0 along the lines $T = \text{constant}$. To perform this calculation, we need $(\partial C_V/\partial V)_T$ and this follows from the equation of state thus:

$$\left(\frac{\partial C_V}{\partial V}\right)_T = \left(\frac{\partial}{\partial V}\right)_T \left(T\frac{\partial S}{\partial T}\right)_V = T\frac{\partial^2 S}{\partial V \partial T} = T\left(\frac{\partial}{\partial T}\right)_V \left(\frac{\partial S}{\partial V}\right)_T = T\left(\frac{\partial^2 P}{\partial T^2}\right)_V \qquad (2.4.1)$$

after making use of equations (2.3.2) and (2.2.12). Integrating the last equation with respect to V, commencing from the initial value $C_V(V_0, T)$, we obtain

$$C_V(V, T) = C_V(V_0, T) + T \int_{V_0}^{V} (\partial^2 P / \partial T^2)_V \, dV. \qquad (2.4.2)$$

Evidently, this defines $C_V(V, T)$ over the same range of values of T for which $C_V(V_0, T)$ is known.

In the case of a Van der Waals gas, substitution from equation (2.3.10) shows that $\partial C_V / \partial V = 0$, i.e. $C_V = C_V(V_0, T)$ and is a function of T alone.

For a gas obeying Berthelot's equation, viz.

$$(P + a/TV^2)(V - b) = vRT, \qquad (2.4.3)$$

we find $\partial C_V / \partial V = -2a/T^2 V^2$ and, thus,

$$C_V(V, T) = C_V(V_0, T) + \frac{2a}{T^2} \left(\frac{1}{V} - \frac{1}{V_0} \right). \qquad (2.4.4)$$

Similarly, we expect $(\partial C_P / \partial P)_T$ to be calculable from the equation of state and find, using equations (2.3.6) and (2.2.13),

$$\left(\frac{\partial C_P}{\partial P} \right)_T = \left(\frac{\partial}{\partial P} \right)_T \left(T \frac{\partial S}{\partial T} \right)_P = T \frac{\partial^2 S}{\partial P \, \partial T} = T \left(\frac{\partial}{\partial T} \right)_P \left(\frac{\partial S}{\partial P} \right)_T = -T \left(\frac{\partial^2 V}{\partial T^2} \right)_P. \qquad (2.4.5)$$

Again, a single integration from the initial value $C_P(P_0, T)$ gives $C_P(P, T)$.

Clearly, C_V and C_P must be related in such a way that, when either is given, the other is calculable by use of the equation of state alone. Suppose we regard S as a function of T and V; the equation of state expresses V as a function of T and P; hence

$$S(T, V) = S\{T, V(T, P)\} = \Sigma(T, P). \qquad (2.4.6)$$

Differentiating this equation partially with respect to T keeping P constant, we find

$$\left(\frac{\partial S}{\partial T} \right)_V + \left(\frac{\partial S}{\partial V} \right)_T \left(\frac{\partial V}{\partial T} \right)_P = \left(\frac{\partial \Sigma}{\partial T} \right)_P = \left(\frac{\partial S}{\partial T} \right)_P, \qquad (2.4.7)$$

where the (strictly incorrect) replacement of Σ by S can safely be made, since the notation employed clearly indicates, for each differentiated state variable, the two variables on which it is being taken to depend. Multiplying the last equation through by T and making use of equations (2.3.2), (2.3.6) and (2.2.12), it can be written

$$C_P - C_V = T \left(\frac{\partial P}{\partial T} \right)_V \left(\frac{\partial V}{\partial T} \right)_P, \qquad (2.4.8)$$

which is the relationship required. For an ideal gas, this yields $C_P - C_V = vR$, as already found at equation (1.5.16).

Other fluid moduli which serve to specify important properties of a fluid are (i) β, the *expansion coefficient*, (ii) k_T, the *isothermal compressibility*, and (iii) k_S, the *adiabatic compressibility*.

β measures the fractional increase in volume per unit rise in temperature at constant pressure. Thus

$$\beta = \frac{1}{V}\left(\frac{\partial V}{\partial T}\right)_P.$$
(2.4.9)

It can be calculated from the equation of state. For an ideal gas, $\beta = 1/T$.

k_T measures the fractional decrease in volume per unit rise in pressure at constant temperature. Thus

$$k_T = -\frac{1}{V}\left(\frac{\partial V}{\partial P}\right)_T.$$
(2.4.10)

For an ideal gas, $k_T = 1/P$.

k_S measures the fractional decrease in volume per unit rise in pressure at constant entropy; i.e. no heat communicated to the fluid, whose contraction is therefore adiabatic. Thus

$$k_S = -\frac{1}{V}\left(\frac{\partial V}{\partial P}\right)_S.$$
(2.4.11)

This coefficient is not calculable from the equation of state directly. It can, however, be determined when $\gamma = C_P/C_V$ and k_T are known, for

$$k_S = k_T/\gamma.$$
(2.4.12)

To prove the last equation, we appeal to the identity

$$\left(\frac{\partial x}{\partial y}\right)_z \left(\frac{\partial y}{\partial z}\right)_x \left(\frac{\partial z}{\partial x}\right)_y = -1,$$
(2.4.13)

which is valid provided any one of the variables x, y, z is a function of the other two. It follows that

$$\left(\frac{\partial S}{\partial T}\right)_P \left(\frac{\partial T}{\partial P}\right)_S \left(\frac{\partial P}{\partial S}\right)_T = -1 = \left(\frac{\partial S}{\partial T}\right)_V \left(\frac{\partial T}{\partial V}\right)_S \left(\frac{\partial V}{\partial S}\right)_T.$$
(2.4.14)

Hence, using equations (2.3.2), (2.3.6), we find

$$\gamma = C_P/C_V = \left(\frac{\partial S}{\partial T}\right)_P \bigg/ \left(\frac{\partial S}{\partial T}\right)_V = \left(\frac{\partial T}{\partial V}\right)_S \left(\frac{\partial V}{\partial S}\right)_T \bigg/ \left\{\left(\frac{\partial T}{\partial P}\right)_S \left(\frac{\partial P}{\partial S}\right)_T\right\}$$

$$= \left(\frac{\partial T}{\partial V} \cdot \frac{\partial P}{\partial T}\right)_S \left(\frac{\partial V}{\partial S} \cdot \frac{\partial S}{\partial P}\right)_T = \left(\frac{\partial P}{\partial V}\right)_S \left(\frac{\partial V}{\partial P}\right)_T = k_T/k_S,$$
(2.4.15)

as stated.

The coefficients introduced thus far refer to reversible changes in the state of the fluid. We next consider two coefficients associated with irreversible processes.

Joule's experiment, in which a gas is permitted to expand into an evacuated chamber without performing work and without gain or loss of heat, has already been mentioned in section 1.5. For an ideal gas, no change of temperature is caused by this process, but for a real gas there is always a small temperature fall.

The *Joule coefficient* μ_J for a fluid measures the temperature drop per unit volume increase due to an infinitesimal Joule expansion process. Since no work is performed on the gas and it is thermally insulated, the internal energy of the gas does not change as a result of the process. Also, although the process is irreversible, for the purpose of calculation it is permissible to assume the equilibrium terminal states are connected by a quasi-static process for which U is constant. Hence,

$$\mu_J = -(\partial T/\partial V)_U. \tag{2.4.16}$$

Then, using the identity (2.4.13), it follows that

$$\mu_J = (\partial T/\partial U)_V (\partial U/\partial V)_T. \tag{2.4.17}$$

But

$$dU = T\,dS - P\,dV = T\{(\partial S/\partial V)_T\,dV + (\partial S/\partial T)_V\,dT\} - P\,dV,$$
$$= \{T\,(\partial P/\partial T)_V - P\}\,dV + T(\partial S/\partial T)_V\,dT, \tag{2.4.18}$$

using equation (2.2.12). Hence

$$(\partial U/\partial V)_T = T(\partial P/\partial T)_V - P. \tag{2.4.19}$$

Also, by equation (1.5.9), $(\partial U/\partial T)_V = C_V$. It now follows from equation (2.4.17) that

$$\mu_J = \{T(\partial P/\partial T)_V - P\}/C_V. \tag{2.4.20}$$

Knowing C_V and the equation of state, this equation permits the calculation of μ_J. For an ideal gas, $\mu_J = 0$ and for a Van der Waals gas $\mu_J = a/(C_V V^2)$.

Next, consider a fluid which is transferred from one cylinder to another via a connecting pipe provided with a tap. Each cylinder is equipped with a frictionless piston to which pressures P_0 and P_1 are applied ($P_0 > P_1$). Suppose initially the high-pressure cylinder contains all the fluid at pressure P_0, volume V_0 and temperature T_0, the low-pressure cylinder being empty. Finally, the high-pressure cylinder is empty and the low-pressure cylinder contains the fluid in the state (P_1, V_1, T_1). If we assume the cylinders are insulated, the increase in the internal energy must equal the work done on the gas, i.e.

$$U_1 - U_0 = P_0 V_0 - P_1 V_1,$$

or

$$U_0 + P_0 V_0 = U_1 + P_1 V_1, \tag{2.4.21}$$

showing that the fluid's enthalpy is unchanged by this irreversible process. This equation, together with the equation of state and the functional dependence of U on V and T, are sufficient to determine T_1, when P_0, P_1 and T_0 are given. In practice, the flow is steadily maintained through the valve and the cooling or heating of the fluid so generated is called the *Joule–Kelvin effect*.

Consider an infinitesimal Joule–Kelvin process in which all changes are supposed very small. The *Joule–Kelvin coefficient* μ_{JK} for the fluid is the fall in temperature per unit drop in pressure, i.e.

$$\mu_{JK} = (\partial T/\partial P)_H. \tag{2.4.22}$$

Again, since the terminal states are being regarded as equilibrium states, the actual process connecting them may be replaced by a quasi-static process. Then, by equation (2.2.7), since H is constant

$$T\,\mathrm{d}S + V\,\mathrm{d}P = 0. \tag{2.4.23}$$

But

$$\mathrm{d}S = (\partial S/\partial T)_P\,\mathrm{d}T + (\partial S/\partial P)_T\,\mathrm{d}P,$$
$$= (C_P/T)\,\mathrm{d}T - (\partial V/\partial T)_P\,\mathrm{d}P, \tag{2.4.24}$$

having applied equations (2.3.6) and (2.2.13). It now follows from the last two equations that

$$C_P\,\mathrm{d}T = \{T(\partial V/\partial T)_P - V\}\,\mathrm{d}P \tag{2.4.25}$$

and therefore that

$$\mu_{JK} = \frac{1}{C_P}\{T(\partial V/\partial T)_P - V\}. \tag{2.4.26}$$

μ_{JK} is zero for an ideal gas, but may be positive or negative for a real gas. For a given pressure, the temperature at which μ_{JK} vanishes for a gas is called its *inversion temperature* for that pressure. In general, there are two inversion temperatures at a given pressure, an upper and a lower; for temperatures between these inversion temperatures, the gas is cooled by a JK-expansion, whereas for temperatures outside this range, it is heated.

2.5 Adiabatic demagnetization

As an alternative example of a two-parameter system, consider a block of magnetic material whose overall magnetic moment is M in the direction of a uniform magnetizing field H. If T is the block's temperature, (H, T) is a complete set of parameters of state (we shall neglect any changes in volume and temperature caused by pressure applied to the block's surface) and the equation of state will be of the form

$$M = M(H, T). \tag{2.5.1}$$

The energy equation has already been found in section 2.1 to be

$$\mathrm{d}U = T\,\mathrm{d}S - \mu_0\,M\,\mathrm{d}H. \tag{2.5.2}$$

A mathematical analysis of the system can now be performed along the lines followed earlier for a fluid, replacing P by M and V by $\mu_0 H$ everywhere.

   In particular, we introduce C_H, the heat capacity of the block with constant field and C_M, the heat capacity with constant moment. These coefficients are given by the formulae (c.f. equations (2.3.2), (2.3.6))

$$C_H = T(\partial S/\partial T)_H, \quad C_M = T(\partial S/\partial T)_M. \tag{2.5.3}$$

Approximate counterparts of k_T and k_S are denoted by α_T and α_S. α_T measures the increase in moment per unit increase in the magnetizing field intensity at constant temperature, i.e.

$$\alpha_T = (\partial M/\partial H)_T \tag{2.5.4}$$

and is termed the *differential isothermal susceptibility* of the block. α_S is the *differential adiabatic susceptibility* of the block and is defined by

$$\alpha_S = (\partial M/\partial H)_S. \tag{2.5.5}$$

As at equation (2.4.15), we can prove that

$$\alpha_T = \Gamma\alpha_S, \tag{2.5.6}$$

where $\Gamma = C_H/C_M$.

An important consequence of the dependence of the magnetization on the temperature is that, under adiabatic conditions, any variation of the magnetic field is accompanied by a change in the block's temperature. Thus, making use of the identity (2.4.13), we find that

$$(\partial T/\partial H)_S = -(\partial S/\partial H)_T(\partial T/\partial S)_H = -\frac{\mu_0 T}{C_H}\left(\frac{\partial M}{\partial T}\right)_H \tag{2.5.7}$$

where, in the last step, we have used the counterpart of the Maxwell relation (2.2.12) and the first of equations (2.5.3). Supposing the material to be paramagnetic and to obey Curie's law, we have $\chi = A/T$ and thus the magnetization $m = AH/T$. Equation (2.1.4) now yields for the moment of the block

$$M = AVH/T, \tag{2.5.8}$$

where V is the volume of the block. This is the state equation for the block. It now follows that

$$(\partial M/\partial T)_H = -AVH/T^2 \tag{2.5.9}$$

and hence, by equation (2.5.7),

$$(\partial T/\partial H)_S = \mu_0 AVH/(C_H T). \tag{2.5.10}$$

This indicates that, as a result of adiabatic demagnetization in which H is reduced, the temperature of the block will drop. To integrate this equation, we need to know C_H as a function of H and T.

Using the equivalent of the Maxwell relation (2.2.12), we have

$$(\partial S/\partial H)_T = \mu_0(\partial M/\partial T)_H = -\mu_0 AVH/T^2 \tag{2.5.11}$$

from equation (2.5.9). Integration with respect to H over the range $(0, H)$ now yields

$$S(H, T) = S(0, T) - \mu_0 AVH^2/(2T^2). \tag{2.5.12}$$

Differentiating this equation with respect to T, keeping H constant, now leads to the result

$$C_H(H, T) = C_H(0, T) + \mu_0 AVH^2/T^2. \tag{2.5.13}$$

At room temperatures and for attainable field strengths, the second term in the right-hand member of this equation is negligible by comparison with the first, i.e. the heat capacity is independent of the field H. Then, equation (2.5.10) can be approximated by the equation

$$C_H(0, T) T \, dT = \mu_0 AVH \, dH. \tag{2.5.14}$$

Suppose that initially the field intensity is H and the temperature is T. If the field is reduced to zero adiabatically, then the temperature drops to T' where, by integration of equation (2.5.14),

$$\int_{T'}^{T} C_H(0, T) T \, dT = \tfrac{1}{2}\mu_0 AVH^2. \tag{2.5.15}$$

The heat capacity per unit volume $C_H(0, T)/V$ will generally be large, so that the temperature fall $T - T' = \Delta T$ will be small. We can then approximate (2.5.15) by $C_H(0, T) T \Delta T = \tfrac{1}{2}\mu_0 AVH^2$, or

$$\Delta T = \frac{\mu_0 AVH^2}{2TC_H(0, T)} \tag{2.5.16}$$

For presently attainable field strengths H, the temperature fall is small.

However, if T is close to absolute zero, it is the second term in the right-hand member of equation (2.5.13) which is dominant and we can approximate this equation by taking

$$C_H(H, T) = \mu_0 AVH^2/T^2. \tag{2.5.17}$$

Now, equation (2.5.10) shows that $dT/T = dH/H$ and, therefore,

$$T = KH, \tag{2.5.18}$$

where K is constant. Thus, according to this approximate equation, if H is reduced to zero, T will also drop to absolute zero. In reality, such *adiabatic demagnetization* can lower temperatures to within a few thousandths of a kelvin of absolute zero, but the limiting temperature cannot be exactly attained.

Indeed, no process has ever been discovered by which any system's temperature can be reduced to $T = 0$ and a third law of thermodynamics is now accepted according to which *by no finite series of processes is absolute zero attainable*. Thus it is possible by a series of processes to approach $T = 0$ as a limit, but the limit itself can never be reached in a finite number of steps (see section 5.5 for further discussion of this law).

2.6 Thermal radiation

It is a familiar circumstance that a hot body placed in a vacuum loses heat energy by radiation. For example, a metal filament heated by the passage of an electric current and placed in an evacuated glass bulb, soon cools when the current is switched off. The energy loss is attributable to the emission of electromagnetic waves from the body, the distribution of the energy flux in this thermal radiation over the frequency spectrum depending upon the nature of the body and its temperature. The inverse phenomenon, of the absorption of electromagnetic radiation incident upon the surface of a body, is also well-known, the ability of the sun's rays to warm objects on which they fall being due to this effect. A body which absorbs all the radiant energy incident upon it will reflect none and, provided we can ignore the thermal radiation it emits due to its own heat, it will appear black under all conditions of illumination. Such an ideal object is therefore called a *blackbody* and the radiation it emits is termed *blackbody radiation*. It will be shown that the distribution of the energy in this radiation over the frequency spectrum is a function of the frequency v and temperature T of the body alone.

Although an ideal blackbody is impossible to construct, the radiation within an evacuated cavity of arbitrary shape inside a block of any opaque material will be shown to have blackbody characteristics. It is assumed that the block and its radiation are in thermodynamical equilibrium and that the walls of the cavity have temperature T. Then, suppose a straight tunnel of small cross-section is bored through the block to the cavity and radiation of any frequency is beamed down the passage. When the ray enters the cavity, it will be repeatedly reflected a large number of times by the walls and, being partially absorbed on each occasion, its energy will ultimately be completely dissipated as heat into the block. Provided the tunnel's cross-section is small, the possibility that a reflected ray can re-enter this channel will be negligible and we are justified in assuming that all the energy in the original incident beam will be absorbed by the cavity. Thus, the cavity will behave like an ideal blackbody and this leads us to expect that the radiation generated within the cavity, which will escape through any borehole, will be characterized by the blackbody spectrum appropriate to the temperature T. This we shall now prove to be the case.

Imagine a small plane area dA to be formed at some point in the cavity and let $f\,dv\,dA$ be the rate of transmission of radiant energy in the frequency band $(v, v + dv)$ across the area in the sense of its unit normal $\mathbf{n}$. Then the radiation intensity f is independent of the point's position, the direction of propagation $\mathbf{n}$, the shape of the cavity and the block material, i.e. $f = f(v, T)$ is a universal function. For, let a pair of differently shaped cavities be excavated in two blocks formed from different materials, both blocks being in thermodynamical equilibrium at the temperature T. Suppose, if possible, points P and Q can be found, one in each cavity, such that the radiation intensity at P for the direction $\mathbf{n}$ and frequency v is greater than the intensity at Q for the direction $\mathbf{m}$ and the same frequency v. Imagine the temperature of the block containing Q is raised by an amount ΔT, so small that the inequality just mentioned is not disturbed. Let a tube of small cross-section made from perfectly reflecting material be arranged to connect P and Q, its axis being aligned with the directions $\mathbf{n}$ and $\mathbf{m}$ at its ends and suppose filters placed at these ends admit into the tube only radiation with frequency in the range $(v, v + dv)$. Then the flux of energy entering the tube at P

will exceed that entering at Q and there will be a net transfer of radiant energy from the cavity at temperature T to that at the higher temperature $T + \Delta T$. This effect would be in violation of Clausius's form of the second law. We conclude, therefore, that f can depend on v and T alone. Thus, the thermal radiation in the cavity is uniform and isotropic. Further, the radiation must be unpolarized, for if it were not we could equip the ends of the tube with Nicol prisms so oriented that the flux of energy into the tube at P exceeded that at Q and the second law would again be breached.

It is assumed in the argument of the previous paragraph that the tube's gauge is sufficiently large to permit the free passage of radiation at the frequency being considered. This means that the dimensions of its cross-section must be at least comparable with the wavelength. For temperatures well above absolute zero, the wavelengths of the significant components of blackbody radiation are very small by ordinary standards (e.g. at 300 K, the dominant wavelength is about 10^{-5} m). This means that the gauge of the tube can be taken sufficiently fine to rule out any appreciable disturbance of the distribution of radiant energy within a cavity having dimensions of the order of a centimetre. However, at 1 K the dominant wavelengths are of the order of a centimetre and, unless the cavity has dimensions which are large compared with this unit, its distribution of radiant energy will be significantly disturbed by the presence of the tube. We cannot therefore conclude that the radiation within a centimetric cavity at this low temperature will be independent of the cavity shape or have blackbody characteristics. At such a low temperature, the cavity would need to be comparatively large to enclose the characteristic blackbody radiation.

Now consider the energy balance at a surface element dS of a cavity. The radiant energy in the frequency band $(v, v + dv)$ incident upon the element per unit time is $f \, dv \, dS$. Suppose a fraction $\alpha (v, T)$ of this energy is absorbed by the surface and converted into heat; then α is called the *absorptance* of the surface for this frequency and temperature. Since the system is in thermodynamical equilibrium, this loss of energy from the cavity must be compensated for by the emission of an equal quantity of radiant energy from dS. Hence, if $e(v, T)$ is the *emissivity* of the surface, i.e. the rate of emission of radiant energy from the surface per unit area and per unit frequency range, then $e \, dv \, dS = \alpha f \, dv \, dS$, or

$$e/\alpha = f. \qquad (2.6.1)$$

We have proved, therefore, that the ratio of the emissivity of a surface to its absorptance is a function of v and T alone. This law was discovered by G. R. Kirchhoff (1824–87). It follows that, since polished metal surfaces are poor absorbers of radiation (most of the radiant energy is reflected), they are also poor emitters; this is the reason the walls of a Dewar's flask are silvered. In the special case of a blackbody surface, $\alpha = 1$ and, hence, $e = f$. This proves finally that cavity radiation is identical with that which is emitted from an ideal blackbody at the same temperature.

To analyse the cavity radiation thermodynamically, we shall treat it as a photon gas and first need to calculate the gas pressure on the walls of the cavity due to photon bombardment, If v is the frequency of a photon, its energy is given by Einstein's relation $E = hv$ (h = Planck's constant = 6.626×10^{-34} J s) and its momentum is E/c, where c is the velocity of light ($c = 2.998 \times 10^8$ m s^{-1}).

At any point in the cavity, let u_v dv denote the energy density of photons whose frequencies lie in the band $(v, v + dv)$; then it has been proved that u_v is a function of T and v alone. Let P be any point on the cavity wall and $\mathbf{n}$ the unit outwards normal from the wall at the point. A small solid angle with apex at P, whose axis makes an acute angle θ with $\mathbf{n}$ will be denoted by dω. Consider all those photons in the cavity whose frequencies lie in the stated range and whose lines of motion are parallel to vectors such as $\mathbf{v}$ (Fig. 2.1a) terminating at P and lying within dω. Since the radiation in the cavity is isotropic, the energy density of these photons is u_v dv d$\omega/4\pi$. Now consider all photons whose lines of motion are parallel to the axis of dω and which impinge upon unit area of the cavity wall in the vicinity of P in unit time. At any instant, these will be contained in an oblique cylinder having volume $c \cos \theta$ (Fig. 2.1b). Photons whose directions of motion lie inside dω and which impinge on unit area at P in unit time will accordingly have energy u_v dv d$\omega c \cos \theta/4\pi$ and their total momentum will therefore be u_v dv d$\omega \cos \theta/4\pi$. It is permissible to assume all these photons will be reflected as indicated in Fig. 2.1b since, even if a photon is absorbed, it must be replaced by the emission of an identical photon along the reflected ray if equilibrium is to be maintained. The momentum change corresponding to the reflection of a photon having momentum p is $2p \cos \theta$ in the direction of $\mathbf{n}$. Thus, the momentum change per unit time caused by the unit area's interception of photons in the stated frequency range and solid angle is u_v dv d$\omega \cos^2\theta/2\pi$ along $\mathbf{n}$. Integrating over all the solid angles dω contained within the hemisphere with its centre at P and lying wholly inside the cavity, we calculate the radiation pressure at P, viz.

$$dP = \frac{u_v \, dv}{2\pi} \int_0^{2\pi} d\phi \int_0^{\frac{1}{2}\pi} \cos^2\theta \sin\theta \, d\theta = \frac{1}{3} u_v \, dv. \qquad (2.6.2)$$

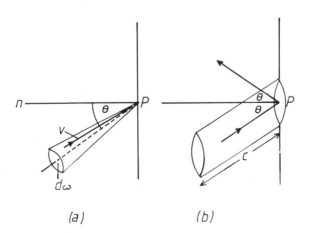

(a) (b)

Fig. 2.1

This is the component of the total pressure P for which photons having frequencies in the range $(v, v + dv)$ are responsible. Integrating over all frequencies, we obtain

$$P = \frac{1}{3} \int u_v \, dv = u/3, \qquad (2.6.3)$$

where u is the overall energy density of the radiation. Clearly u, and therefore P, is a function of T alone.

We now write the energy equation for the photon gas in the standard form $dU = T \, dS - P \, dV$. Putting $U = uV$, $P = u/3$, this gives

$$u \, dV + V \frac{du}{dT} dT = T \, dS - \frac{1}{3} u \, dV, \qquad (2.6.4)$$

or, after rearrangement,

$$dS = \frac{V}{T} \frac{du}{dT} dT + \frac{4u}{3T} dV. \qquad (2.6.5)$$

Since dS is a perfect differential, this implies that

$$\frac{\partial}{\partial V} \left(\frac{V}{T} \frac{du}{dT} \right) = \frac{\partial}{\partial T} \left(\frac{4u}{3T} \right). \qquad (2.6.6)$$

Differentiating out and rearranging, we find

$$du/u = 4 \, dT/T \qquad (2.6.7)$$

which, upon integration, yields *Stefan's law* for blackbody radiation, viz.

$$u = aT^4, \qquad (2.6.8)$$

where a is a universal constant. Its value can be determined experimentally or, using the methods of statistical mechanics, be shown to depend on the constants c, k and h. In SI units, its value is $7.565 \times 10^{-16} \, \text{J} \, \text{m}^{-3} \, \text{K}^{-4}$.

S now follows from equation (2.6.5) in the form

$$S = \frac{4}{3} aVT^3 + \text{constant.} \qquad (2.6.9)$$

The principles of classical thermodynamics alone are insufficient to permit us to determine the manner in which the energy density u_v depends on the radiation frequency v. However, in section 7.6 we shall employ the more powerful methods of statistical mechanics to complete the analysis of blackbody radiation.

2.7 Phase equilibrium

Consider a composite system, each of whose N subsystems is the same homogeneous substance in different phases (e.g. water as ice, liquid and vapour). Initially, the subsystems will be supposed separated from one another by flexible and heat-permeable membranes which ensure that, in equilibrium, the temperature and pressure are equalized over the whole system. The system will be further supposed immersed in a heat bath which maintains its temperature at T and also to be exposed to an ambient (e.g. atmospheric) pressure P. Each subsystem will be taken to behave as a two-parameter system, whose equilibrium state is determined by its temperature and pressure alone. The mass of the rth phase will be denoted by m_r and $\Sigma m_r = M$ will denote the total mass. If the membranes are removed (a process assumed to involve no work), the phases will, in general, be transformed into one another in such a way that the masses of some will be reduced whilst the masses of others will be increased, subject to the condition that M remains constant. However, it will be assumed that the various phases retain their identities, i.e. that there is no mixing.

When all phases are in equilibrium under pressure P and at temperature T, suppose the membranes are briefly removed so that interaction between phases can take place. Immediately afterwards, the membranes are again placed in position and the phase changes are again inhibited. If, during this short interval, heat dQ is supplied to the system from the heat bath and the system's volume increases by dV, then the increase in the internal energy is given by

$$dU = dQ - P\,dV. \tag{2.7.1}$$

During this process, the system is not in equilibrium and consequently the process is irreversible; thus, according to the inequality (1.9.3), we must have $dQ < T\,dS$, where dS is the entropy change. Hence

$$dU - T\,dS + P\,dV < 0. \tag{2.7.2}$$

Since there is no mixing, the entropy of the system is the sum of the entropies of the phases and the Gibbs function for the system is also the sum of the Gibbs functions for the phases. Thus

$$G = U - TS + PV \tag{2.7.3}$$

and, for any process during which T and P are kept constant,

$$dG = dU - T\,dS + P\,dV. \tag{2.7.4}$$

Hence, for the process just described,

$$dG < 0, \tag{2.7.5}$$

i.e. G must decrease.

Let $g_r = g_r(T, P)$ be the Gibbs function per unit mass of the rth phase. Then

$$G = \sum m_r g_r, \quad M = \sum m_r. \tag{2.7.6}$$

When the membranes are removed, since T and P are being held constant, the functions g_r will not vary and the m_r will change in such a way that G decreases. If we suppose the coefficient g_R has a value which is less than all the rest, then G takes its minimum value Mg_R when all the masses m_r vanish, except $m_R = M$. In these circumstances, repeated removal of the membranes will ultimately result in the

44

disappearance of all phases except the Rth, i.e. it is impossible for the various phases to remain in equilibrium with one another. If, however, the g_r all have equal values g, then $G = Mg = $ constant and no irreversible change of the type we have assumed can occur. In this case, all the phases remain in equilibrium in the presence of one another after the removal of the membranes.

There are, of course, intermediate cases. If $g_1 = g_2$ and the remaining g_r all take larger values, the system will ultimately settle into an equilibrium state in which only the phases 1 and 2 survive. The masses of these phases will be indeterminate except for the condition $m_1 + m_2 = M$; any two masses satisfying this condition will represent a possible equilibrium state and the actual state selected will depend upon the values imposed on other state variables, e.g. the volume.

Consider, for example, a substance (such as water) which has solid, liquid and gaseous phases, for which the Gibbs functions per unit mass are g_S, g_L, g_G respectively. In the PT-plane, we plot the three curves

$$\phi_{SL} = g_S - g_L = 0, \quad \phi_{LG} = g_L - g_G = 0, \quad \phi_{GS} = g_G - g_S = 0, \quad (2.7.7)$$

which, in the case of water, take the forms sketched in Fig. 2.2 (*phase diagram*). Each curve $\phi = 0$ separates the PT-plane into two regions, in one of which $\phi > 0$ and in the other $\phi < 0$. Thus, in the region below the curve $\phi_{GS} = 0$, we find $g_G < g_S$ and, in the region above this curve $g_G > g_S$. Similarly, below the curve $\phi_{LG} = 0$, we find $g_G < g_L$, and above, $g_G > g_L$; to the left of the curve $\phi_{SL} = 0$, we have $g_S < g_L$, and to the right, $g_S > g_L$. The reader will now be able to verify that the three curves define six regions, which have been labelled from 1 to 6, and that, in these regions, the inequalities listed below are valid:

Region 1: $g_G < g_S < g_L$.
Region 2: $g_G < g_L < g_S$.
Region 3: $g_L < g_G < g_S$.
Region 4: $g_L < g_S < g_G$.
Region 5: $g_S < g_L < g_G$.
Region 6: $g_S < g_G < g_L$.

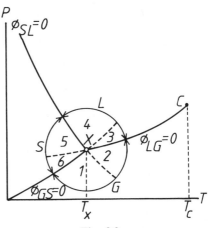

Fig. 2.2

It now follows from what has been proved that, for temperatures and pressures in regions 1 and 2, the coefficient g_G is smallest and the substance will exist in its gaseous phase. In regions 3 and 4, g_L is smallest and the substance will be liquid. Finally, in regions 5 and 6, g_S is smallest and the substance will be solid. On that part of the curve $\phi_{LG} = 0$ separating regions 2 and 3 (continuous line), we have $g_L = g_G < g_S$ and the liquid and gaseous phases can coexist, but the solid phase is prohibited; this is termed the *vapour-pressure curve*. On the part of the curve $\phi_{SL} = 0$ separating regions 4 and 5, we have $g_S = g_L < g_G$ and the liquid and solid phases can coexist, with the gaseous phase being prohibited; this is termed the *melting curve*. Finally, on the curve $\phi_{GS} = 0$ separating regions 1 and 6, $g_S = g_G < g_L$ and the solid and gaseous phases can coexist, with the liquid phase prohibited; this is the *sublimation curve*. The three curves meet at the *triple point* X, where $g_S = g_L = g_G$ and all three phases can coexist in equilibrium.

Having proved that, for most ambient temperatures and pressures, only one phase of a substance is viable, some justification is needed for our earlier assumption that, prior to the removal of the partitioning membranes, all the phases are present within the composite system. In fact, it has been demonstrated experimentally that most liquids can be cooled below their melting points without solidifying, provided they are free from impurities which might provide nuclei around which solidification can commence and are shielded from contact with the substance in its solid phase; such liquids are then said to be *supercooled*. Further, by slowly raising the temperature of a pure liquid uniformly in a vibration-free container, its boiling point can be exceeded without transition to the gaseous phase; it is then *superheated*. This property of liquid hydrogen is employed in *bubble chambers* to track particles generated by nuclear reactions; when pressure on the hydrogen is reduced, it becomes superheated, but its boiling is inhibited except in the neighbourhood of the particle tracks where ionized hydrogen atoms act as centres around which bubbles of hydrogen gas are formed. Vapours, also, can be supercooled below their temperatures of condensation to the liquid phase; again, dust, or other particles, may provide nuclei around which droplets form and this phenomenon has been used in the *Wilson cloud chamber* to make visible the ionized tracks of atomic particles.

These anomalous states are said to be *metastable* and are not true equilibrium states. Nevertheless, the possibility of their existence indicates that our hypothesis of the initial presence in the composite system of all the phases of a substance, provided these are denied contact with one another, is not totally non-physical, especially if the ambient conditions are not far from those at the triple point. To justify our calculation of the equation of the vapour-pressure curve, we need only accept the possibility that superheated liquid can exist in a narrow band of states on the vapour side of the curve and that supercooled vapour can exist in a similar narrow band on the liquid side, thus establishing the existence of both the functions g_L and g_G in these regions.

Referring to Fig. 2.2, at a point (P, T) on the melting curve we have $g_S = g_L$. Suppose we displace to a point $(P, T + \delta T)$ just to the right of this curve and let g_S increase by δg_S and g_L increase by δg_L as a result. Now, in region 4, $g_S > g_L$ and it follows that $\delta g_S > \delta g_L$. We conclude that $\delta g_S/\delta T > \delta g_L/\delta T$ or, in the limit on the melting curve

$$s_S = -(\partial g_S/\partial T)_P < -(\partial g_L/\partial T)_P = s_L, \qquad (2.7.8)$$

where s is the entropy per unit mass (use equation (2.2.11)). Thus, the transition from the solid to the liquid phase involves an increase in entropy and this implies that heat must be supplied to the substance to effect the transformation, i.e. the *latent heat of fusion* is positive. Similarly, we can prove that the *latent heats of vaporization* and *of sublimation* are positive.

Suppose, instead, we move from a point (P, T) on the melting curve to a neighbouring point $(P + \delta P, T)$ above the curve in region 4. Then $\delta g_S/\delta P > \delta g_L/\delta P$ and in the limit, therefore,

$$v_S = (\partial g_S/\partial P)_T > (\partial g_L/\partial P)_T = v_L, \qquad (2.7.9)$$

where v is the specific volume. This inequality shows that melting ice contracts, explaining why ice floats on water. For most substances, however, the melting curve has positive gradient and the reader should verify that, if Fig. 2.2 is amended accordingly, a displacement from the melting curve in the direction of increasing pressure brings us into region 5 instead of 4 and the inequality (2.7.9) is reversed. Thus, most substances expand on melting. Transitions from the liquid or solid phases to the gaseous phase always result in expansion of the substance as, indeed, follows from Fig. 2.2.

2.8 Vapour-pressure curve. Clapeyron's equation

We next obtain an explicit equation for the vapour-pressure curve by treating the vapour as a perfect gas and assuming that changes in the volume of the liquid phase can be neglected.

For an ideal gas, $U = C_V T$ and equation (1.7.6) gives the entropy. Thus, the Gibbs function for the gas is

$$G = C_V T - C_V T \ln T - \nu RT \ln V + PV + AT, \qquad (2.8.1)$$

where A is the unknown additive constant associated with the entropy. Making use of the equation of state $PV = \nu RT$ and the equation (1.5.16) for C_P, the Gibbs function can be expressed as a function of P and T thus:

$$G = \nu RT \ln P - C_P T \ln T + BT, \qquad (2.8.2)$$

where B is a new constant. This is the Gibbs function for ν mols of gas; the Gibbs function per mol is given by

$$g_G = RT \ln P - c_p T \ln T + \beta T, \qquad (2.8.3)$$

where $\beta = B/\nu$ and $c_p = C_P/\nu$ is the molar heat capacity at constant pressure.

For an ideal liquid, we assume $C_P = C_V = C(T)$. If, therefore, we neglect any variation in V, we have for a reversible process, $dU = C\,dT = T\,dS$ and, hence,

$$U = \int C\,dT, \qquad (2.8.4)$$

$$S = \int (C/T)\, dT. \tag{2.8.5}$$

It now follows from equation (2.2.6) that

$$G = \int C\, dT + PV_L - T \int (C/T)\, dT. \tag{2.8.6}$$

Writing $c = C/v$ for the molar heat capacity of the liquid,

$$g_L = \int c\, dT + Pv_L - T \int (c/T)\, dT, \tag{2.8.7}$$

where v_L is the molar volume of the liquid phase.

Referring to equations (2.8.3) and (2.8.7), we can now write down the equation $g_L = g_G$ of the vapour-pressure curve in the form

$$\ln P = \frac{v_L}{RT} P + (c_P/R)\ln T + (1/RT) \int c\, dT - (1/R) \int (c/T)\, dT, \tag{2.8.8}$$

after absorbing the constant β in the second indefinite integral. The term $v_L P/RT$ is always small by comparison with the others and may be neglected to yield an explicit expression for P in terms of T.

Since $g = h - Ts$, on the vapour-pressure curve we must have

$$h_G - h_L = T(s_G - s_L). \tag{2.8.9}$$

But $(s_G - s_L)$ is the increase in entropy when a mol of liquid at temperature T is transformed into a mol of vapour at the same temperature. Hence, $T(s_G - s_L)$ is the heat which must be supplied during this process and this is the *molar latent heat of vaporization* l. Using the results

$$h_G = u_G + Pv_G = c_V T + RT = c_P T, \quad h_L = \int c\, dT + Pv_L, \tag{2.8.10}$$

equation (2.8.9) yields the result

$$l = c_P T - \int c\, dT - Pv_L. \tag{2.8.11}$$

Now differentiate equation (2.8.8) with respect to T; this gives

$$\left(\frac{1}{P} - \frac{v_L}{RT}\right)\frac{dP}{dT} = \frac{1}{RT^2}\left(c_P T - \int c\, dT - v_L P\right) = \frac{l}{RT^2}. \tag{2.8.12}$$

But, if v_G is the molar volume of the vapour at temperature T and pressure P, then $Pv_G = RT$ and the last equation accordingly reduces to

$$\frac{dP}{dT} = \frac{l}{T(v_G - v_L)} \tag{2.8.13}$$

This is *Clapeyron's equation*. Since l is always positive and $v_G \gg v_L$, this equation proves that P is an increasing function of T along the vapour-pressure curve.

This last equation can also be derived by differentiation with respect to T of the vapour-pressure equation in the form $g_L(P,T) = g_G(P,T)$ to give

$$\left(\frac{\partial g_L}{\partial P}\right)_T \frac{\mathrm{d}P}{\mathrm{d}T} + \left(\frac{\partial g_L}{\partial T}\right)_P = \left(\frac{\partial g_G}{\partial P}\right)_T \frac{\mathrm{d}P}{\mathrm{d}T} + \left(\frac{\partial g_G}{\partial T}\right)_P. \tag{2.8.14}$$

From equation (2.2.11), it follows that $(\partial g_L/\partial P)_T = v_L$, $(\partial g_G/\partial P)_T = v_G$, $(\partial g_L/\partial T)_P = -s_L$ and $(\partial g_G/\partial T)_P = -s_G$. Hence

$$\frac{\mathrm{d}P}{\mathrm{d}T} = \frac{\Delta s}{\Delta v}, \tag{2.8.15}$$

where $\Delta s = s_G - s_L$ is the entropy increase and $\Delta v = v_G - v_L$ is the volume increase when one mol of liquid is transformed into a mol of vapour at temperature T. Since $l = T\Delta s$, the last equation can be written

$$\frac{\mathrm{d}P}{\mathrm{d}T} = \frac{l}{T\Delta v}, \tag{2.8.16}$$

which is Clapeyron's equation again. Note that Clapeyron's equation remains valid if l is taken to be the latent heat per unit mass and v the volume per unit mass.

Suppose the liquid is contained in a cylinder kept at temperature T and fitted with an airtight piston which, initially, rests on the liquid surface. If the piston is withdrawn from contact with the liquid, the space so created above the liquid will be filled with vapour and the pressure exerted upon the underside of the piston by this vapour will be P as given by equation (2.8.8). If the piston is now moved towards the liquid, the vapour pressure will not change, sufficient vapour condensing to the liquid phase to keep P constant. If the piston is removed from the cylinder, air will mix with the vapour so that, in equilibrium, the sum of the partial pressures of the air and vapour is equal to the atmospheric pressure. If the temperature T is raised, the vapour pressure will ultimately equal the atmospheric pressure and the air's partial pressure in the cylinder will be reduced to zero (i.e. no air will enter). Any further rise in temperature will cause the vapour pressure to exceed the atmospheric and equilibrium cannot then be attained; at this point the liquid boils and will be steadily transformed into vapour until the ambient pressure is raised. Since we have proved that P increases with T, the temperature for boiling always increases with the pressure (it is well known that water boils at a lower temperature at the top of a mountain than it does at sea level).

After consideration of the manner in which Clapeyron's equation (2.8.16) was derived, it will be evident that it remains valid on the melting and sublimation curves. Since water contracts on melting, Δv will be negative and the equation shows that the slope of the melting curve will be negative, as we have already remarked. Increasing the pressure lowers the melting point of ice and thus, by applying pressure to the ice, a skater creates a film of water which serves to lubricate the passage of the skate's blade as it moves over the ice. When a solid sublimates, Δv is always positive and the sublimation curve therefore always has a positive gradient.

A final characteristic of the phase diagram (Fig. 2.2) which we shall note, is that the vapour-pressure curve terminates at a point C, termed the *critical point*. If the temperature of the substance is maintained constant at a value between the critical point temperature T_C and the triple point temperature T_X, a steadily increasing pressure upon the gaseous phase will eventually bring the substance's state to a point on the vapour-pressure curve. The vapour will now commence to condense and, when it has been completely transformed into liquid, further increase in the pressure compresses this liquid until the melting curve is reached (if ever), when a change to the solid phase takes place. If, however, the temperature is kept above the critical temperature T_C, no intersection with the vapour-pressure curve occurs and there is no phase change until a point on the melting curve is reached. At such a temperature, there is no distinction between the liquid and gaseous phases, changes of pressure serving only to vary the degree of liquidity or gaseousness of the substance. Thus, even, at high pressure, when the substance behaves substantially as a liquid, if it is held in a container under gravity, its density varies continuously with height and no distinct liquid surface develops at any pressure. For water, the critical temperature is 374 °C and the pressure at the critical point is 219 atmospheres.

2.9   Gibbs' phase rule

Suppose a composite system comprises K substances, each able to exist in a number (not, necessarily, all the same) of phases. It will be assumed that the substances do not react chemically. In some metastable state (membranes in position), let $m_i^{(k)}$ be the mass of the ith phase of the kth substance. Then, if $M^{(k)}$ is the total mass of the kth substance,

$$m_1^{(k)} + m_2^{(k)} + \ldots = M^{(k)}, \quad k = 1, 2, \ldots, K. \tag{2.9.1}$$

If $g_i^{(k)}$ is the Gibbs function per unit mass of the ith phase of the kth substance, the Gibbs function for the metastable state is

$$G = \sum_{k=1}^{K} (m_1^{(k)} g_1^{(k)} + m_2^{(k)} g_2^{(k)} + \ldots). \tag{2.9.2}$$

In the true equilibrium state, G will be minimized under variation of the $m_i^{(k)}$ subject to the constraints (2.9.1). For the kth substance, this means that all coefficients $m_i^{(k)}$ become zero except the one associated with the $g_i^{(k)}$ having minimum value amongst the set $\{g_1^{(k)}, g_2^{(k)}, \ldots\}$. However, it is possible a number of members of this set have equal minimum value, in which case all their coefficients can be non-zero and the corresponding phases can all survive. Suppose

$$g_1^{(k)} = g_2^{(k)} = \ldots = g_{i_k}^{(k)} \tag{2.9.3}$$

is the set of minimum $g_i^{(k)}$ for the kth substance. (Note: since each $g_i^{(k)}$ is a function of only two variables P and T, i_k cannot, in general, exceed 3, for otherwise we would have more conditions than unknowns.) The number of conditions provided by the equations (2.9.3) is $i_k - 1$ and the total number for all the substances will therefore be

$$\sum_{k=1}^{K} (i_k - 1) = I - K, \qquad (2.9.4)$$

where $I = \Sigma i_k$ is the total number of distinct phases present in the equilibrium state. Only two variables P and T are available to satisfy these conditions and, accordingly, we must have $I - K \leqslant 2$, or

$$I \leqslant K + 2. \qquad (2.9.5)$$

This is *Gibbs' phase rule*.

If $K = 1$, then $I \leqslant 3$ and it is possible for up to three phases to be present in an equilibrium state. This is the triple point already noted.

If $K = 2$, then $I \leqslant 4$ and it is possible for four phases to coexist at three *quadruple points*: (i) three phases of substance A and one of B, (ii) three phases of B and one of A, (iii) two phases of each of A and B.

Exercises 2

1. Writing the energy equation for a fluid in the form

$$dS = \frac{1}{T} dU + \frac{P}{T} dV$$

and regarding S as a function of U and V, calculate $\partial^2 S / \partial U \, \partial V$ in two different ways to show that

$$\left(\frac{\partial T}{\partial V} \right)_U = P \left(\frac{\partial T}{\partial U} \right)_V - T \left(\frac{\partial P}{\partial U} \right)_V.$$

2. Solving the fluid energy equation for dV prove, as in the previous exercise, that

$$\left(\frac{\partial P}{\partial S} \right)_U = P \left(\frac{\partial T}{\partial U} \right)_S - T \left(\frac{\partial P}{\partial U} \right)_S.$$

3. Writing the energy equation of a fluid in the form

$$dS = \frac{1}{T} dH - \frac{V}{T} dP,$$

prove that

$$\left(\frac{\partial T}{\partial P} \right)_H = T \left(\frac{\partial V}{\partial H} \right)_P - V \left(\frac{\partial T}{\partial H} \right)_P.$$

4. Prove

$$\text{(i)} \quad \left(\frac{\partial U}{\partial V}\right)_T = -P + T\left(\frac{\partial P}{\partial T}\right)_V,$$

$$\text{(ii)} \quad \left(\frac{\partial H}{\partial P}\right)_T = V - T\left(\frac{\partial V}{\partial T}\right)_P.$$

5. For any fluid, prove that

(i) $(\partial U/\partial S)_V = (\partial H/\partial S)_P = T$, (ii) $(\partial S/\partial P)_H = -V/T$,

$$\text{(iii)} \quad \left(\frac{\partial}{\partial U}\frac{P}{T}\right)_V = \left(\frac{\partial}{\partial V}\frac{1}{T}\right)_U.$$

6. From the equation

$$dS = (\partial S/\partial P)_V \, dP + (\partial S/\partial V)_P \, dV$$

deduce that

$$(\partial P/\partial V)_S = -(\partial S/\partial V)_P \, (\partial P/\partial S)_V.$$

Show, similarly, that

$$(\partial P/\partial V)_T = -(\partial T/\partial V)_P \, (\partial P/\partial T)_V$$

and hence prove that

$$\frac{(\partial P/\partial V)_S}{(\partial P/\partial V)_T} = \gamma.$$

7. Prove the results

$$\text{(i)} \quad (\partial T/\partial P)_S = -(\partial S/\partial P)_T \, (\partial T/\partial S)_P = \frac{T}{C_P}\left(\frac{\partial V}{\partial T}\right)_P,$$

$$\text{(ii)} \quad (\partial T/\partial V)_S = -\frac{T}{C_V}\left(\frac{\partial P}{\partial T}\right)_V.$$

8. Using equation (2.3.2) and Maxwell's relation (2.2.3), prove

$$C_V = -T(\partial V/\partial T)_S \, (\partial P/\partial T)_V.$$

Using the identity (2.4.13), deduce that

$$C_V = -\frac{T\beta}{k_T}\left(\frac{\partial V}{\partial T}\right)_S.$$

9. Using equation (2.4.8), prove that

$$\gamma = 1 + \frac{\beta^2 TV}{C_V k_T}.$$

10. Defining the coefficient of tension τ for a fluid by the equation

$$\tau = \frac{1}{P}\left(\frac{\partial P}{\partial T}\right)_V,$$

use the identity (2.4.13) to prove that $Pk_T\tau = \beta$.

11. Show that, for a Van der Waals gas, the inversion temperature T at pressure P is determined by the equation

$$(3\tau + 2\pi - 6)^2 + 16(\tau + 2\pi - 2) = 0,$$

where $\tau = bvRT/a$, $\pi = b^2P/a$. Show that the inversion curve plotted from this equation in the $\pi\tau$-plane is a parabola and identify the regions of the plane where the gas is heated or cooled by a JK-expansion.

12. Dieterici's equation of state for a gas is

$$P(V - b) = vRT \exp\{-a/(vRTV)\}.$$

Show that the inversion curve has equation

$$\pi = (1 - \tau)\exp(1 - 1/\tau)$$

where τ and π are as defined in the previous exercise. Sketch this curve and identify the heating and cooling regions for a JK-expansion. Show that, if $\pi > \frac{1}{2}(3 - \sqrt{5})\exp\{-\frac{1}{2}(\sqrt{5} - 1)\}$, a JK-expansion invariably heats the gas.

13. Prove for a two-parameter magnetic system that

$$(\partial U/\partial M)_T = \mu_0\{H - T(\partial H/\partial T)_M\}.$$

Deduce that, if the substance is paramagnetic and obeys Curie's law, then U depends upon T alone.

14. Prove that $C_M = (\partial U/\partial T)_M$ for a magnetic system with two parameters. Deduce that, if the system obeys Curie's law, then C_M depends on T alone.

15. Assuming the molar latent heat l is constant in Clapeyron's equation (2.8.13) and that the vapour obeys the ideal gas law $Pv = RT$, obtain the following equation fot the vapour-pressure curve if v_L can be ignored by comparison with v_G:

$$P = P_0 \exp\left\{\frac{l}{R}\left(\frac{1}{T_0} - \frac{1}{T}\right)\right\}.$$

16. Given that, at $0\,°C$, the molar latent heat of fusion for water is 6.03×10^3 J/mol, the molar volume of ice is 1.9633×10^{-5} m^3/mol and the molar volume of water is 1.8002×10^{-5} m^3/mol, calculate the pressure in atmospheres needed to reduce the melting point by $1\,°C$. (1 atm $= 1.013 \times 10^5$ N m^{-2}). (Ans. 135 atm.)

17. Given that, at $100\,°C$, the latent heat of vaporization of water is 4.063×10^5 J/mol, the molar volume of water is 1.877×10^{-5} m^3/mol and the molar volume of steam is 0.3011 m^3/mol, calculate the boiling point of water at the top of a mountain where the atmospheric pressure is 0.5 atm. (Ans. $86\,°C$.)

18. The equation of state for a certain gas is Callendar's equation

$$V = \frac{RT}{P} + b - c/T^n,$$

where b, c are positive constants and n is a positive integer. At low pressures, the gas behaves ideally and $C_P = a$ (constant). Show that

$$C_P = n(n+1)cP/T^{n+1} + a,$$

$$S = a \ln T - R \ln P - cnP/T^{n+1}.$$

CHAPTER 3

Classical statistics.
Maxwell's distribution

3.1 Hypothesis of random states

Statistical mechanics provides a mathematical method for predicting the behaviour of any physical system whose state we are unable to specify with the
maximum precision permitted by the fundamental theories of mechanics (e.g.
classical, relativistic or quantum mechanics, whichever is being employed as a
basis for our calculations). Such a lack of possible information in regard to the
state of a system is usually due to the system's complexity, which renders it
impractical to list the values of the multitude of observable quantities which
would need to be known to determine completely the system's state at some initial
instant and to permit an exact calculation of its subsequent evolution by
application of the accepted mechanical principles. Thus, the systems we shall be
studying will normally comprise a number of virtually autonomous subsystems,
each belonging to one of a small number of categories, the subsystems within each
category possessing identical structures; the number of such subsystems will be so
vast, that a full account of the state of each cannot be given. Further, even if it were
possible to provide an exhaustive survey of the states of the subsystems at some
instant, the complexity of their subsequent interactions would prohibit a
successful calculation of their behaviour, even though this is, in principle,
deducible from the operation of known laws of mechanics. The type of system we
have in mind is a mixture of gases confined in a container, in which case the
subsystems are the gas molecules and the number of types of subsystems is the
number of different gases forming the mixture. Another example is a sample of
some crystalline substance whose subsystems will be the atoms arranged in the
crystal lattice.

Nevertheless, our general method will not be restricted to assemblies of
virtually autonomous subsystems, but will be applicable to provide information
regarding the behaviour of any system, simple or complex, integrated or
separable, for which a complete specification of state cannot, for one reason or
another, be provided. Clearly, such information cannot be expected to be precise,
but will be expressed in terms of the probabilities of the system being observed to
exhibit various characteristics; i.e. the data provided will be statistical in character.
However, in the special case of an assembly of a very large number of subsystems,
it will be shown that the probability distributions of all the physical quantities of

interest are very sharply peaked, so that the values of these likely to be observed can be predicted with great confidence and experimental verification of the theory can proceed as though its forecasts were exact. The most impressive successes of statistical mechanics have been achieved in connection with this type of system, but its methods are available for treating quite simple systems, such as individual molecules (e.g. their Brownian motion), in which cases the imprecisions, which are a characteristic feature of any statistical theory, play more important roles.

The concept of the probability a given system occupies a certain state at a certain time must accordingly be of fundamental importance for the theory we are about to develop. But such a concept cannot be derived from others which make no reference to the measurement of the chance that certain events will, in given circumstances, occur. It is necessary, at the outset, to secure agreement regarding the probabilities which are to be assigned, in stated circumstances, to the occurrence of members of some basic set of events, which can then act as the starting point for the theoretical development. Thus, it may be accepted that the probabilities that an unbiased coin when spun should fall heads or tails are both one-half. That this is the correct assignment of probabilities to these events cannot be proved by argument from purely non-probabilistic features of the system being considered, for it is evident that this allocation of probabilities is only correct provided the coin is unbiassed and that the coin has this property can only be demonstrated by tossing it a very large number of times and verifying that the numbers of heads and tails do not deviate appreciably from equality. Thus, the probabilities arising in any statistical theory must ultimately be sanctioned by empirical evidence—either the probabilities assigned to the basic events need to be checked experimentally by repeated generation of these events or, alternatively, probabilities of other events, as predicted by the theory, must be subjected to such experimental check. For example, if it is assumed that ten coins are unbiassed, instead of checking each individually, the coins may be spun together and the number of heads counted on each occasion—the frequencies of 0, 1, 2, etc. heads are predicted by probability theory and can be compared with the experimental data; substantial agreement would confirm the hypothesis that the coins are unbiassed.

The most common way of choosing the fundamental set of events to which probabilities are allocated is to identify a set which, in given circumstances, it seems reasonable to regard as equally probable. We then say that we are supposing the events to occur *at random*. As explained above, to be acceptable our supposition would require direct experimental verification or its consequences would have to be shown to be in conformity with our experience. In section 3.3, we shall establish a theorem of classical mechanics which indicates how we might choose a set of states for a classical system which it is reasonable to accept, on the basis of the limited information available, are equally probable. This choice will then be employed as the foundation on which the classical theory of statistical mechanics is built, the ultimate justification for adopting the hypothesis being that the theoretical predictions are in accord with the experimental facts. In so far as there are discrepancies, it has been found that these can be removed, not by amending the hypothesis in regard to the random distribution of states, but by replacing the classical ideas by those of quantum mechanics; in section 5.3, we shall derive the appropriate assumption in regard to the random distribution of quantum states upon which to found the theory of quantum statistics.

3.2 The Gibbs ensemble

Using the generally accepted notation of classical analytical mechanics, let $q_i(i = 1, 2, \ldots, n)$ be generalized coordinates for the system under consideration and let p_i denote the associated generalized components of momentum. In most cases with which we shall be concerned, the number n of degrees of freedom of the system will be very large, but at this stage we do not stipulate that this is necessarily so. If $H(q, p, t)$ is the system's Hamiltonian, its motion from a given initial state $q = q^0, p = p^0$ at $t = t^0$ is determined by the Hamilton equations

$$\dot{q}_i = \frac{\partial H}{\partial p_i}, \qquad \dot{p}_i = -\frac{\partial H}{\partial q_i}. \tag{3.2.1}$$

The system will always be assumed conservative, so that H will not depend explicitly on t and the equations of motion will possess a well-known first integral

$$H = \text{constant}, \tag{3.2.2}$$

H being the energy of the system, which is therefore conserved.

It is helpful to regard $(q_i, \ldots, q_n, p_1, \ldots, p_n)$ as rectangular Cartesian coordinates of a point in a $2n$-dimensional Euclidean space called the *phase space*. Since a complete specification of the dynamical state of our system can be given at any time by listing the values of its coordinates and momenta, all possible states of the system are represented on a 1–1 basis by the points of the phase space. Further, as the motion proceeds, the q_i and p_i will vary as functions of t and the representative point will trace out a curve in phase space which we shall refer to as a *phase trajectory* for the system. Clearly, there is a unique trajectory through each point of phase space and the family of trajectories represents all possible motions of the system.

As explained in the previous section, the precise state of the type of system we shall be studying will never be known and instead we shall hope to assign probabilities to the various possible states which are in conformity with the constraints imposed by the limited knowledge we do possess (e.g. the system's energy may be known, in which case the probability of a state of different energy is zero). Having identified the probability of an event with the relative frequency with which the event occurs in a very large number of trials, it is convenient to model the possible states of our system at some instant t^0, together with their associated probabilities, by imagining that a very large number G of copies of the system are prepared, such that the number of these copies in a state S^0 is Gp, where p is the probability for the state S^0. Such a model of this probabilistic situation is termed an *ensemble* and was introduced into the theory by the American mathematician J. W. Gibbs (1839–1903); clearly, the frequency with which a particular state is found in the ensemble is proportional to its associated probability. It is understood, of course, that the copies are isolated from one another, so that they do not interact, and each is subjected to the same external influences. The behaviour of each copy is then completely determined by the equations of motion (3.2.1) and, if the ensemble is properly constructed at some initial instant t^0, its overall state at some later time t will correctly fix the probabilities of the various possible states at this instant. This must be so since, if p is the relative frequency with which the system is observed in the state S^0 at time t^0, then p must also be the relative frequency with which the system will be observed in the corresponding evolved state S at a later time t.

At any instant, the state of a member of the ensemble will be represented by a point in phase space and the state of the whole ensemble can therefore be specified by a set of G points in the space. Since G is very large, we can picture the phase space as being occupied by a fluid comprising point molecules and define the density of the fluid ρ at any point (q, p) at time t as the number of points per unit volume of the space in the neighbourhood of the point at this instant. Thus, $\rho = \rho(q, p, t)$ and the number of points with coordinates in the ranges $(q, q + dq)$, $(p, p + dp)$ at time t, is $\rho \, dq_1 \ldots dq_n \, dp_1 \ldots dp_n = \rho \, dq \, dp$ (clearly, we are not supposing the differentials to be so small that the discrete structure of the fluid becomes significant). By integration over the whole of the phase space, we must arrive at the result

$$\int \rho \, dq \, dp = G. \tag{3.2.3}$$

By definition of the ensemble, the probability that our actual system at time t has coordinates and momenta lying in the ranges $(q, \; q + dq)$, $(p, \; p + dp)$ respectively is $\rho \, dq \, dp/G$. Thus to determine how a system's probability distribution over its states changes with time, we need to calculate the function $\rho(q, p, t)$; a method for doing this is given in the next section.

3.3 Liouville's theorem

As the time t increases, each point representing the state of a member of the Gibbs ensemble describes a trajectory in phase space and the point fluid streams from one region to another. During this flow, the individual points are conserved and there will be an equation satisfied by ρ expressing this principle. If **v** is the velocity of flow of a fluid whose density is ρ, it is proved in texts devoted to fluid mechanics that the conservation law is expressed by the equation

$$\frac{\partial \rho}{\partial t} + \text{div } \rho \mathbf{v} = 0. \tag{3.3.1}$$

The argument leading to this equation is easily amended to apply to our $2n$-dimensional phase space and since the velocity components of a phase point are

58

$(\dot{q}, \dot{p})$, we arrive at the equation

$$\frac{\partial \rho}{\partial t} + \sum_{i=1}^{n} \left\{ \frac{\partial}{\partial q_i}(\rho \dot{q}_i) + \frac{\partial}{\partial p_i}(\rho \dot{p}_i) \right\} = 0. \tag{3.3.2}$$

Substituting for $\dot{q}_i$, $\dot{p}_i$ from equations (3.2.1), we find

$$\frac{\partial \rho}{\partial t} + \sum_{i=1}^{n} \left\{ \frac{\partial}{\partial q_i}\left(\rho \frac{\partial H}{\partial p_i} \right) - \frac{\partial}{\partial p_i}\left(\rho \frac{\partial H}{\partial q_i} \right) \right\} = 0, \tag{3.3.3}$$

which reduces to

$$\frac{\partial \rho}{\partial t} + \sum_{i=1}^{n} \left(\frac{\partial \rho}{\partial q_i}\frac{\partial H}{\partial p_i} - \frac{\partial \rho}{\partial p_i}\frac{\partial H}{\partial q_i} \right) = 0. \tag{3.3.4}$$

The sum appearing in this equation is called the *Poisson bracket* for ρ and H and is denoted by $\{\rho, H\}$. Thus, we have finally

$$\frac{\partial \rho}{\partial t} = -\{\rho, H\} = \{H, \rho\}. \tag{3.3.5}$$

This is *Liouville's theorem*.

If the distribution of ρ over the whole of phase space is known at an initial instant t^0, so that $\rho(q, p, t^0)$ is a known function of the qs and ps then the first-order partial differential equation (3.3.4) determines $\rho(q, p, t)$ uniquely for all q_i, p_i and $t > t^0$.

There are a number of special cases of this result to which we now direct attention. Firstly, suppose that initially we have no knowledge whatsoever relating to the system's state. Thus we have no reason to prefer one set of values of the coordinates q_i and momenta p_i to any other; the initial probability distribution of the system over its states is completely random. It follows that, unless there is a bias inherent within classical mechanics (e.g. that momenta always tend towards small, or large, values), all values of the quantities (q, p) will have equal probabilities and $\rho = \rho_0$ (constant) at $t = t^0$. This is the *uniform ensemble*. Equation (3.3.4) then has the obvious solution $\rho = \rho_0$ for $t > t^0$, satisfying the stated initial condition. This result implies that the ensemble remains uniform with passage of time and hence that the probability distribution over states continues to be random. It must be our expectation that complete initial ignorance in regard to the state of a system will continue as time elapses and the conclusion we have reached is accordingly perfectly acceptable; it should be seen as confirming our assumption that the coordinate and momentum values are free from bias. It should be noted that this conclusion is not a trivial one, since it is only valid provided the system's state is fixed by giving values of coordinates and momenta; this very fundamental *principle of randomicity in phase*, which we have established, would not be valid if the Hamilton specification of states was abandoned in favour of the Lagrange determination by coordinates q_i and velocities $\dot{q}_i$.

In cases when $\rho(q, p, t)$ is independent of t, we say that the ensemble is in *statistical equilibrium.* The uniform ensemble is one such case, but there are many others. Thus, suppose Hamilton's equations (3.2.1) possess a first integral

$$\alpha(q, p) = \text{constant}. \tag{3.3.6}$$

Differentiation with respect to t gives

$$\sum_{i=1}^{n} \left(\frac{\partial \alpha}{\partial q_i} \dot{q}_i + \frac{\partial \alpha}{\partial p_i} \dot{p}_i \right) = 0, \tag{3.3.7}$$

or

$$\sum_{i=1}^{n} \left(\frac{\partial \alpha}{\partial q_i} \frac{\partial H}{\partial p_i} - \frac{\partial \alpha}{\partial p_i} \frac{\partial H}{\partial q_i} \right) = 0. \tag{3.3.8}$$

It therefore follows that if we take ρ to be any function of α, thus:

$$\rho = f(\alpha) = \rho(q, p), \tag{3.3.9}$$

upon substitution in equation (3.3.4) we get

$$0 + \sum_{i=1}^{n} \frac{df}{d\alpha} \left(\frac{\partial \alpha}{\partial q_i} \frac{\partial H}{\partial p_i} - \frac{\partial \alpha}{\partial p_i} \frac{\partial H}{\partial q_i} \right) = 0 \tag{3.3.10}$$

and equation (3.3.8) shows that Liouville's equation is satisfied. But $\rho = f(\alpha)$ is independent of t and so defines an ensemble in statistical equilibrium.

In particular, assuming the system is conservative, the energy integral (3.2.2) is available and we can take $\alpha = H$; then $\rho = f(H)$ defines an ensemble in statistical equilibrium. For a system modelled by such an ensemble, the probability for any state depends only on the state's energy, states of equal energy being equally probable; because of the equilibrium, this distribution of probabilities does not change with time. A special case is when the function $f(H)$ is zero, except for values of the energy confined to a narrow range $(E, E + \delta E)$; then, the probability the system has an energy outside this range is zero and the ensemble models a system whose energy is known within narrow limits. Over the range of states having the required energy, the probability distribution for the system is uniform and therefore purely random, in accordance with the principle of randomicity in phase. This is called the *microcanonical ensemble.*

3.4   Thermodynamical equilibrium

Since we shall usually be concerned with systems having a very large number n of degrees of freedom (e.g. millions of molecules rather than one or two), Liouville's equation (3.3.4) will normally involve a large number of terms and ρ will also depend upon a large number of variables. It is then impossibly difficult to find the solution of the equation satisfying a given initial condition. For example, we might wish to study a system in the form of a container separated into two compartments, the one filled with air and the other evacuated. At time t^0, the wall separating the compartments is removed and it is required to discuss, by statistical methods, the process by which the gas expands into the vacuum. Assuming that the gas was originally ($t < t^0$) in statistical equilibrium in its compartment, it is not difficult to model the initial state of the system by the appropriate ensemble— clearly ρ will vanish at all points in phase space corresponding to points in the empty compartment. However, the evolution of this ensemble as determined by Liouville's equation would be quite impossible to calculate. We must accordingly restrict our attention to systems in statistical equilibrium for which solutions of the equation are available. Fortunately, an ensemble in statistical equilibrium is the appropriate model for a system in thermodynamical equilibrium, for our knowledge regarding the state of such a system does not change as time elapses. Thus, restricting ourselves to this type of ensemble is equivalent to confining our attention to physical systems in thermodynamical equilibrium, a limitation which has already been accepted in the development of classical theory and which is recognized to leave ample scope for numerous useful applications.

In this connection, it is of interest to remark that Boltzmann and Gibbs have constructed mathematical arguments which show that, if we construct an ensemble to represent the initial state of an isolated assemblage of a large multitude of similar interacting molecules, it can be predicted with a high degree of probability that, after some lapse of time (usually short), the state of this ensemble will be indistinguishable from that of a microcanonical ensemble. This means that any such ensemble ultimately achieves statistical equilibrium and, at the same time, the system of molecules it represents arrives at a state of thermodynamical equilibrium. This provides justification for modelling an isolated system of molecules in thermodynamical equilibrium by a microcanonical ensemble, as we shall do in the next section.

For a system in statistical equilibrium, the mean values of all physical quantities associated with the system as a whole or with comparatively large elements of the system (such as gas pressure and density) will be independent of t and, due to the peaked nature of the probability distributions for such quantities (see section 3.1), this implies that the observed values of these quantities will also be constants. Such quantities will be described as *macroscopic* to distinguish them from *microscopic* observables such as the coordinates and momenta of the individual molecules, whose distributions have a wider spread.

The precise physical significance of the mean value of any quantity (e.g. the system's energy) calculated over an ensemble in statistical equilibrium has been the subject of controversy. The natural attitude of the statistician is to argue that it is a measure of the mean value of observations of the quantity made upon a large number of systems, all in states which have been prepared by identical procedures. However, another point of view is that a single system is prepared and observations of the quantity are then taken at a large number of successive instants, from which the mean is calculated. The assumption that the means calculated by these two distinct procedures will be identical is called the *ergodic hypothesis*. A proof of the hypothesis could be constructed if it could be shown that, over a long period of time, the system's state lies in each neighbourhood of phase space for an interval proportional to the phase density ρ. Unfortunately, this cannot be done, since the system's equations of motion will possess first integrals corresponding to various principles of conservation (e.g. angular momentum) and the existence of these constraints will exclude the phase trajectory from some regions of the phase space where ρ fails to vanish. Again, it is possible that the system selected for observation over time could be in some quite exceptional state, so that observations made upon it could not be regarded as representative of these systems in general circumstances (e.g. all the molecules of a gas in a smooth-walled box might be moving parallel to one pair of faces indefinitely); in such exceptional circumstances, the two means would certainly differ. We prefer, therefore, to abandon any attempt to identify a mean calculated over an ensemble with a time-average for a single system and to accept the orthodox statistical interpretation.

3.5 Maxwell–Boltzmann distribution law

As an example of a system which can be modelled by a microcanonical ensemble, consider a gas composed of N identical molecules confined to a perfectly rigid container isolating it from the environment so that its total energy is conserved with value U. The gas will be supposed so dilute that the molecules interact very rarely and the energy of interaction can be neglected. However, at this stage, we shall not exclude action upon the molecules by an external conservative field such as gravity. The gas will be supposed to have settled into statistical equilibrium, so that the observed values of all macroscopic quantities like the gas density or pressure are constants.

If the dynamical state of the μth molecule is specified in terms of coordinates $q_r^{(\mu)}$ and momenta $p_r^{(\mu)}$ ($\mu = 1, 2, \ldots, N; r = 1, 2, \ldots, n$), the dynamical state of the whole gas is determined by the set of $2Nn$ quantities $(q_r^{(\mu)}, p_r^{(\mu)})$; thus, the gas system has Nn degrees of freedom and its phase space has $2Nn$ dimensions. A phase space can also be identified for each individual molecule—it will evidently have $2n$ dimensions. In order that combinatorial methods can be employed, it is now necessary to imagine that the molecular phase spaces are each divided into a very large number of cells, in the shape of rectangular boxes with edges parallel to the q and p axes; the corners of these boxes will form a lattice extending over the whole of the molecular phase space. Since all products $p_r q_r$ have the same physical dimensions (energy × time = action), it is convenient to assume that the product of the lengths of the pair of cell's edges parallel to corresponding p and q axes is always equal to a small quantity h; then, the volume of a cell will be h^n. The cells will be supposed numbered according to some convenient scheme (the same for all spaces) so that, h being small, the state of a molecule is almost precisely specified by nominating the cell number to which its representative state point belongs. This subdivision of the molecular phase spaces will induce a division of the gas phase space into cells of volume h^{Nn}. However, instead of fixing the gas state by nomination of one of these cells, it is more convenient to give the numbers a_1, a_2, etc. of gas molecules whose individual states lie in the cells 1, 2, etc. of the molecular phase spaces (since the cells in these spaces are numbered in the same way, henceforward we can merge them into a single all-purpose structure). This procedure fails to distinguish between all those distinct states of the gas obtained by permutation of the molecules between cells without variation of the numbers $a_1, a_2, \ldots$; but this is an advantage, since the macroscopic observables will depend only on the numbers $a_1, a_2, \ldots$ and not at all on the particular molecules allocated to the various cells. We shall say that such a specification identifies a *condition* of the gas, reserving the term 'state' for the more precise determination.

Not all the cells, either in molecular phase space or in gas phase space, can be regarded as available for occupation. Only those corresponding to gas states for which the molecules lie within the container and have total energy in some range $(U, U + \delta U)$ are open. But, having identified the open cells in the gas phase space, by the principle of randomicity in phase, the probability the gas state lies in one of these cells must be the same for all. Thus, the same number of members of our ensemble must be allocated to each open cell; since we intend to let $h \to 0$, so that the number of cells becomes very large, and each cell shrinks to a point, it is convenient to let each cell contain a single member of the ensemble, i.e. we take the number G of members of the ensemble to be the same as the number of admissible cells in gas phase space.

The number of distinct states of a gas embraced by the condition $(a_1, a_2, \ldots)$ described above is the number of ways N objects can be allocated, a_1 to the first cell, a_2 to the second cell, and so on; this number is well-known to be $N!/a_1!\, a_2! \ldots$. Each of these distinct states of the gas is modelled by the single member of the microcanonical ensemble in the corresponding cell in the gas phase space and, hence, the number of members of the ensemble modelling the condition $(a_1, a_2, \ldots)$ is

$$W = \frac{N!}{a_1!\, a_2! \ldots} \tag{3.5.1}$$

and the probability for this condition is accordingly

$$p = \frac{N!}{a_1!\, a_2! \ldots} \cdot \frac{1}{G}. \tag{3.5.2}$$

Clearly, the numbers a_i must satisfy the constraint

$$a_1 + a_2 + \ldots = N. \tag{3.5.3}$$

Also, if ε_i is the total energy (kinetic and potential) for a molecule whose state is represented by a point in the ith cell of the molecular phase space (it is reasonable to ignore any variation of energy over the cell since $h \to 0$), then it is necessary that

$$a_1\, \varepsilon_1 + a_2\, \varepsilon_2 + \ldots = U. \tag{3.5.4}$$

Thus, the numbers a_i must satisfy the two constraints (3.5.3) and (3.5.4).

The natural next step would be to analyse the probability distribution (3.5.2) over the range of possible gas conditions and to demonstrate that, if N is large, this distribution is sharply peaked, so that mean or expected values of quantities calculated over the distribution can be identified with observed values of these quantities. If, however, we assume this result for the moment, a short cut to some important conclusions can be made by calculating the admissible gas condition for which p is a maximum; for a sharply peaked distribution this, also, can be identified with the observed condition.

If N is large, in the most probable condition it seems reasonable to suppose a very high proportion of the a_i will be sufficiently large to permit approximation of their factorials by Stirling's formula (see Appendix B), thus

$$a! = \sqrt{(2\pi)} e^{-a} a^{a+\frac{1}{2}}. \tag{3.5.5}$$

Use of this result in equation (3.5.2) yields

$$\ln p = \sum_i \left\{ a_i - \left(a_i + \frac{1}{2} \right) \ln a_i \right\} + K, \tag{3.5.6}$$

where K is independent of the a_i. Using condition (3.5.3) and neglecting the terms $\ln a_i$ by comparison with the terms $a_i \ln a_i$, the last equation reduces to

$$\ln p = - \sum_i a_i \ln a_i + \text{constant}. \tag{3.5.7}$$

Thus, we need to minimize $\sum a_i \ln a_i$ subject to the constraints (3.5.3) and (3.5.4). Introducing Lagrange multipliers α and β, we must minimize the function

$$\phi = \sum a_i \ln a_i + \alpha \sum a_i + \beta \sum a_i \varepsilon_i, \tag{3.5.8}$$

with all constraints relaxed. Treating the a_i as continuous real variables, differentiation yields

$$\frac{\partial \phi}{\partial a_i} = 1 + \ln a_i + \alpha + \beta \varepsilon_i = 0. \tag{3.5.9}$$

Thus,

$$a_i = \exp(-1 - \alpha - \beta \varepsilon_i) = A e^{-\beta \varepsilon_i}. \tag{3.5.10}$$

This is the *Maxwell–Boltzmann law* for the most probable distribution of the gas molecules amongst their various energy states (Note: the cell energies ε_i are not necessarily all different).

The constants A, β are determined by the two constraints

$$\sum_i A e^{-\beta \varepsilon_i} = N, \qquad \sum_i A \varepsilon_i e^{-\beta \varepsilon_i} = U. \tag{3.5.11}$$

Thus, β is fixed by the equation

$$\left(\sum \varepsilon_i e^{-\beta \varepsilon_i} \right) \bigg/ \left(\sum e^{-\beta \varepsilon_i} \right) = U/N \tag{3.5.12}$$

and then A follows from

$$A = N \bigg/ \left(\sum e^{-\beta \varepsilon_i} \right). \tag{3.5.13}$$

Writing

$$Z = \sum_i e^{-\beta \varepsilon_i}, \tag{3.5.14}$$

Z is termed the *partition function* for a molecule of the gas and equations (3.5.12), (3.5.13) can then be written

$$\frac{\partial}{\partial \beta}(N \ln Z) = -U, \qquad A = N/Z. \tag{3.5.15}$$

Also note that equation (3.5.10) can be expressed as

$$a_i = -\frac{1}{\beta}\frac{\partial}{\partial \varepsilon_i}(N \ln Z). \tag{3.5.16}$$

3.6 Thermodynamics of a Maxwell gas

Since $U = \sum a_i \varepsilon_i$ (equation (3.5.4)), the energy of the gas can be changed in two ways, (i) by varying the a_i and (ii) by altering the ε_i. Suppose both sets of quantities are given small increments, so that a_i increases by da_i and ε_i increases by $d\varepsilon_i$, these changes being brought about quasi-statically so that the gas never departs from its state of equilibrium. Then the resulting increment in U is given by

$$dU = \sum a_i\,d\varepsilon_i + \sum \varepsilon_i\,da_i. \tag{3.6.1}$$

The changes in the cells' energies contribute the first term in this expression and these can only be brought about by varying the external fields. Thus, the gas molecules will be repelled from the walls of the container by a short-range field of force, whose effect will be to reverse the normal component of the velocity of any molecule incident upon it. This field of force could be subject to variation by arranging for one of the walls to be a piston moving in a cylinder in which the gas is confined and displacing the piston slightly. Let $\theta_1, \theta_2, \ldots$ be geometrical parameters determining the configurations of the external fields (e.g. the distance of the piston from some datum position might be one such parameter), so that the ε_i are functions of the θ_r. Then, if θ_r changes to $\theta_r + d\theta_r$, the resulting energy increment for the gas is

$$\sum_i a_i\,d\varepsilon_i = \sum_{i,r} a_i(\partial \varepsilon_i/\partial \theta_r)d\theta_r = -\sum_r F_r\,d\theta_r,^* \tag{3.6.2}$$

where $F_r = -\sum_i a_i\,\partial \varepsilon_i/\partial \theta_r$. This result shows that the external fields can be regarded as exerting forces on the gas, the resultant of which has generalized component $-F_r$ corresponding to the generalized field coordinate θ_r; the energy increment is then the work done by these forces upon the gas. By Newton's third law, F_r will be the rth component of force exerted by the gas on the external field (e.g. if V is the volume of gas trapped in a cylinder by a piston and P is the gas pressure, $P\,dV$ is the work done on the piston by the gas when V increases by dV

---

* Subsequently, the a_i will adjust to new values $a_i + \delta a_i$, but during this process energy is conserved; thus $\sum \varepsilon_i \delta a_i = 0$ and the second term in (3.6.1) is unaffected.

(section 1.5); thus P is the force component corresponding to the generalized coordinate V).

The second term in the expression (3.6.1) for dU cannot be so easily interpreted and neither can the increments da_i upon which it depends be generated as easily as the increments dε_i. Evidently, the isolation of the gas from its environment will need to be broken if a general rearrangement of molecules among the cells of molecular phase space is to be effected. This is described as placing the gas in contact with a heat source (or sink), by which heat energy d$Q = \Sigma \varepsilon_i \, \mathrm{d}a_i$ is transferred to (or from) the assembly of molecules.

Thus, the internal energy of a gas can be increased by performing work on it via the external force fields and by supplying it with heat energy from an external reservoir. If dQ is the heat energy supplied in the process, then

$$\mathrm{d}Q = \sum \varepsilon_i \, \mathrm{d}a_i = \mathrm{d}U - \sum a_i \, \mathrm{d}\varepsilon_i. \tag{3.6.3}$$

Putting $\psi = N \ln Z = \psi(\beta, \varepsilon_1, \varepsilon_2, \ldots)$ and assuming β increases to $\beta + \mathrm{d}\beta$ as a result of the process, we find by differentiation that

$$\mathrm{d}\psi = (\partial \psi / \partial \beta)\mathrm{d}\beta + \sum (\partial \psi / \partial \varepsilon_i)\mathrm{d}\varepsilon_i = -U \, \mathrm{d}\beta - \beta \sum a_i \mathrm{d}\varepsilon_i, \tag{3.6.4}$$

using equations (3.5.15), (3.5.16). It now follows that

$$\mathrm{d}(\psi + \beta U) = \beta \mathrm{d}U - \beta \sum a_i \mathrm{d}\varepsilon_i = \beta \mathrm{d}Q. \tag{3.6.5}$$

But, as explained in section 1.7, Q is not defined as a function of the gas state, whereas $\psi + \beta U$ is such a function. Thus, equation (3.6.5) shows that dQ is transformed into an exact differential by the factor β. Now, it is known from classical thermodynamics that, for a process of the type we have been considering, dQ/T is an exact differential denoted by dS, where S is the entropy of the gas and T is its absolute temperature. Hence, concordance between statistical and thermo-dynamical theory will be established if we take

$$\beta = 1/kT, \qquad S = k(\psi + \beta U) = \Psi + U/T, \tag{3.6.6}$$

where k is a constant and

$$\Psi = k\psi = kN \ln Z. \tag{3.6.7}$$

k is called *Boltzmann's constant* and its value in SI units is 1.38066×10^{-23} J K^{-1}. The constant β has therefore been identified as a measure of gas temperature and, by comparing equations (2.2.5) and (3.6.6), we also see that

$$\Psi = -F/T, \tag{3.6.8}$$

where F is the free energy. (Note: An arbitrary constant could be added to S in equation (3.6.6), but it is shown in section 5.5 that this can be taken to be zero.)

Further, since $\varepsilon_i = \varepsilon_i(\theta_1, \theta_2, \ldots)$, we have $\psi = \psi(\beta, \varepsilon_1, \varepsilon_2, \ldots) = \psi(\beta, \theta_1, \theta_2, \ldots)$ and hence

$$\mathrm{d}\psi = \sum_i (\partial \psi / \partial \varepsilon_i)\mathrm{d}\varepsilon_i + (\partial \psi / \partial \beta)\mathrm{d}\beta = \sum_r (\partial \psi / \partial \theta_r)\mathrm{d}\theta_r + (\partial \psi / \partial \beta)\mathrm{d}\beta. \tag{3.6.9}$$

66

Cancellation gives

$$\sum_i (\partial\psi/\partial\varepsilon_i)d\varepsilon_i = \sum_r (\partial\psi/\partial\theta_r)d\theta_r, \tag{3.6.10}$$

and then, using equations (3.5.16) and (3.6.2), this leads to

$$\sum_r F_r d\theta_r = \frac{1}{\beta}\sum_r (\partial\psi/\partial\theta_r)d\theta_r. \tag{3.6.11}$$

Since the $d\theta_r$ are independent increments, it follows that

$$F_r = \frac{1}{\beta}\frac{\partial\psi}{\partial\theta_r} = T(\partial\Psi/\partial\theta_r). \tag{3.6.12}$$

In particular, if V is the volume of the container and V increases to $V + dV$, the work done by the gas is $P\,dV$, where P is the gas pressure. Hence

$$P = T(\partial\Psi/\partial V) = kNT\frac{\partial}{\partial V}(\ln Z). \tag{3.6.13}$$

Thus we have derived the second of equations (2.2.15) for the gas.

For the purposes of further analysis, it will be convenient to express the Maxwell–Boltzmann law in differential, instead of discrete, form. Take any small volume element $d\mu$ of the molecular phase space in the vicinity of the ith cell and let dn be the number of molecules with states embraced by the element. Then $dn \propto d\mu$ and equation (3.5.10) can be written

$$dn = Ce^{-\beta\varepsilon}d\mu, \tag{3.6.14}$$

where ε is the energy for these molecules and C is a constant. Since the total number of molecules is N, C must be given by

$$C\int e^{-\beta\varepsilon}d\mu = N, \tag{3.6.15}$$

the integration being carried out over the whole of the molecular phase space.

3.7 Gas of point molecules

In this section, we consider the special case of a gas whose molecules can be regarded as point masses having kinetic energy due to their translatory motion alone. If we further ignore any external field, the energy of a molecule will be given by the equation

$$\varepsilon = \frac{1}{2m}(p_x^2 + p_y^2 + p_z^2), \tag{3.7.1}$$

where m is the molecular mass and (p_x, p_y, p_z) are the Cartesian components of the linear momentum. Coordinates in molecular phase space are (x, y, z, p_x, p_y, p_z) and equation (3.6.14) therefore gives

$$dn = C \exp\left\{ -\frac{\beta}{2m}(p_x^2 + p_y^2 + p_z^2) \right\} dx\, dy\, dz\, dp_x\, dp_y\, dp_z, \qquad (3.7.2)$$

for the number of molecules in the vicinity of (x, y, z) with linear momenta in the neighbourhood of (p_x, p_y, p_z). Clearly, the distribution of molecules over the interior of the container is uniform. Integrating over this volume V and putting $p_x = mc_x$, etc., we find that the number of molecules with velocities close to (c_x, c_y, c_z) is given by

$$dn = CVm^3 \exp\{ -\tfrac{1}{2}\beta m(c_x^2 + c_y^2 + c_z^2) \}\, dc_x\, dc_y\, dc_z. \qquad (3.7.3)$$

Finally, integrating over the whole of velocity space and using the identity

$$\int_{-\infty}^{\infty} \exp(-\alpha x^2)dx = (\pi/\alpha)^{1/2}, \qquad (3.7.4)$$

we find that

$$N = \int dn = CV(2\pi m/\beta)^{3/2}. \qquad (3.7.5)$$

Thus

$$C = \frac{N}{V}(\beta/2\pi m)^{3/2} = \frac{N}{V}(2\pi mkT)^{-3/2}. \qquad (3.7.6)$$

Equation (3.7.3) is *Maxwell's distribution law for velocities*.

To calculate the partition function Z given by equation (3.5.14), we first write

$$Z = h^{-3} \sum \exp(-\beta\varepsilon)h^3, \qquad (3.7.7)$$

where ε is given by equation (3.7.1) and the summation is over the relevant cells of the molecular phase space. Since h^3 is the volume of such a cell, if h is taken very small

$$Z = h^{-3} \int \exp(-\beta\varepsilon)\, dx\, dy\, dz\, dp_x\, dp_y\, dp_z, \qquad (3.7.8)$$

where the integration is over the appropriate region of the phase space. The integration with respect to the Cartesian coordinates (x, y, z) yields a factor V, and thus

$$Z = h^{-3} Vm^3 \int \exp(-\beta\varepsilon)dc_x\, dc_y\, dc_z = h^{-3} V(2\pi m/\beta)^{3/2}, \qquad (3.7.9)$$

having used the identity (3.7.4). Substitution in the first of equations (3.5.15) now gives

$$U = 3N/2\beta = \frac{3}{2}kNT \qquad (3.7.10)$$

as the relationship between the temperature and the internal energy. Note that, as already deduced for an ideal gas (section 1.8, problem 5), U is a function of T alone and, by equation (1.5.10), the heat capacity is given by

$$C_V = \frac{3}{2} kN. \tag{3.7.11}$$

Also, substituting for Z into equation (3.6.13), we derive the perfect gas law, viz. $PV = kNT$ (see equation 1.5.4)). It follows immediately that, if P and T are given, then $N \propto V$, i.e. equal volumes of gases at a given temperature and pressure contain the same number of molecules (*Avogadro's hypothesis*).

Using equations (3.6.6) and (3.6.7), the entropy of the gas is found to be given by

$$S = kN \left(\frac{3}{2} \ln T + \ln V + C \right), \tag{3.7.12}$$

where

$$C = \frac{3}{2} \{ \ln(2\pi mk/h^2) + 1 \}. \tag{3.7.13}$$

Since $kN = \nu R$ (see section 1.5) and C_V is given by equation (3.7.11), equation (3.7.12) is in agreement with the thermodynamcially derived equation (1.7.6).

Unfortunately, equation (3.7.12) cannot be regarded as satisfactory for, using the gas law, it is seen to be equivalent to

$$S = kN \left(\frac{5}{2} \ln T - \ln P + \ln N + C' \right), \tag{3.7.14}$$

where $C' = C + \ln k$. We now see that, if the volume of gas under consideration is enlarged by increasing N without change in the temperature or pressure, S does not increase in proportion to N as it ought, owing to the presence of the term $kN \ln N$. Before the theory which has been developed can be accepted, therefore, some justification is needed for the elimination of this term from equation (3.7.14). It will be shown in section 5.8 that the indistinguishability of the gas molecules coupled with a basic principle of quantum mechanics provides just such a sanction and thus removes a blemish from the theory.

If a conservative external field of force (such as gravity) is to be allowed for, let $W(x, y, z)$ be the potential energy of a molecule at (x, y, z) in the field and replace equation (3.7.1) by

$$\varepsilon = \frac{1}{2m} (p_x^2 + p_y^2 + p_z^2) + W. \tag{3.7.15}$$

Then, equation (3.7.2) has to be amended to read

$$dn = C \exp \{ -\tfrac{1}{2} \beta m (c_x^2 + c_y^2 + c_z^2) + W \} dx\, dy\, dz\, dc_x\, dc_y\, dc_z. \tag{3.7.16}$$

Integrating over the whole of velocity space using the identity (3.7.4), this leads to the formula

$$dn = B e^{-\beta W} dV \tag{3.7.17}$$

for the number of molecules in a volume element $dV = dx\,dy\,dz$ of ordinary space.

The density ρ of the gas is now found to be given by

$$\rho = m\,dn/dV = \rho_0 e^{-\beta W}, \tag{3.7.18}$$

where $\rho_0 = B$ is the density on the equipotential $W = 0$ (e.g. ground level in the case of gravity).

For a uniform gravitational field, we can take $W = mgz$, where z is the height above ground, and then

$$\rho = \rho_0 e^{-\beta mgz} = \rho_0 e^{-\gamma z/T}, \tag{3.7.19}$$

a familiar result.

3.8 Boltzmann's principle. Disorder of systems

The number of equally probable states W modelling a condition of the Maxwell gas is given by equation (3.5.1). W is termed the *statistical weight* of the condition. We have identified the observed condition of the gas with that for which W takes its maximum value, the occupation numbers a_i for this condition being given by equation (3.5.10). We will now calculate W_{max}.

Using Stirling's approximation (3.5.5), we find from equation (3.5.1) that

$$\ln W = \ln \sqrt{(2\pi)} - N + (N + \tfrac{1}{2})\ln N - \sum_i \{\ln \sqrt{(2\pi)} - a_i + (a_i + \tfrac{1}{2})\ln a_i\}. \tag{3.8.1}$$

Neglecting the relatively small terms $\ln N$ and $\ln a_i$, and using the constraint (3.5.3), this reduces to

$$\ln W = N \ln N - \sum_i a_i \ln a_i + \text{constant}, \tag{3.8.2}$$

the constant depending only on the number of cells in the molecular phase space and, therefore, being independent of the gas.

Substituting for the a_i from equation (3.5.10), we now find

$$\ln W_{max} = N \ln N - A \sum_i e^{-\beta \varepsilon_i}(\ln A - \beta \varepsilon_i) + \text{const.} \tag{3.8.3}$$

This can be simplified with the aid of equations (3.5.11) to give

$$\ln W_{max} = N \ln N - N \ln A + \beta U + \text{const.} \tag{3.8.4}$$

Further, since $N/A = Z$ (equation (3.5.15)), we conclude that

$$\ln W_{max} = N \ln Z + \beta U + \text{const.} \tag{3.8.5}$$

Finally, referring to equations (3.6.6), (3.6.7), we see that this implies

$$S = k \ln W_{max} + C. \tag{3.8.6}$$

Thus, the entropy of the gas has been expressed in terms of the statistical weight of its equilibrium condition.

Thus far, the entropy has only been defined for equilibrium states. However, the concept of statistical weight can be extended to molecular systems which are not in thermodynamical equilibrium. Thus, we can define W to be the number of equally probable microstates of the system which model a specific non-equilibrium condition of the system and then define the entropy of the condition by the equation

$$S = k \ln W + C. \qquad (3.8.7)$$

This is *Boltzmann's principle*.

We next note that conditions which are highly ordered tend to have low statistical weights, whereas conditions which are disordered or chaotic tend to have high statistical weights. Thus, if the atoms of a system are known to be arranged in a crystal lattice, the number of microstates modelling the system for a given energy will be far smaller than if the atoms are free to move as a gas throughout a container of the same volume as the crystal.

Or again, consider a gas confined to one compartment of a container provided with a partition. If the partition is removed, the gas will very rapidly reach a new equilibrium state in which it occupies both compartments. The condition of the gas can be regarded as more ordered before the partition is removed than afterwards, for the positions of its molecules are known with greater exactitude then. Also, the number of microstates modelling its original equilibrium state will be smaller than the number modelling its final condition, since the smaller volume will limit the number of molecular phase space cells available for occupation.

As a third example, take a bar magnet assembled from a large number of molecular dipoles. An extreme condition of the system, generating maximum resultant magnetic moment of the bar, would be for all the dipoles to be aligned; only one microstate could model this condition, i.e. $W = 1$. Other conditions, yielding smaller bar moments, in which the configuration of dipoles was less orderly, could be modelled in many ways and would have larger statistical weights.

Such considerations as these suggest that W can be interpreted as a measure of the disorder of a system. If this is accepted, Boltzmann's principle indicates that the entropy is also a measure of disorder and the law of increasing entropy (section 1.9) can then be expressed as requiring that thermally isolated systems must always evolve so as to increase their disorder. Since interactions between the constituent molecules will proceed in a fortuitous manner, this tendency for any initial ordered structure to be destroyed and superseded by disorder (in so far as this is permitted by the nature of the system) is easily understandable. This, then, has now been revealed as the essential content of the second law of thermodynamics.

As an illustration of these ideas, suppose a crystal comprises N atoms, each of which can occupy either of two sites with energies E and $E + \varepsilon$. In a particular macrostate of the crystal, suppose n atoms occupy the higher energy site. The number of distinguishable microstates which can model this macrostate is the number of ways of choosing these n higher energy sites from the total number N of sites, viz. $N!/\{n!(N-n)!\}$. This, then, is the statistical weight of the macrostate. Approximating $\ln(n!)$ by $n(\ln n - 1)$ (assuming n large), equation (3.8.7) gives for the entropy

$$S = -k\{n \ln n + (N-n) \ln(N-n)\} + \text{constant}. \qquad (3.8.8)$$

The internal energy of the macrostate is

$$U = n(E+\varepsilon) + (N-n)E = n\varepsilon + NE. \qquad (3.8.9)$$

We shall assume that variation of n does not affect the volume of the crystal. Then, by the first of equations (2.2.2), we deduce that the crystal's temperature T is given by

$$T = \frac{\partial U}{\partial S} = \left(\frac{\partial U}{\partial n}\right) \bigg/ \left(\frac{\partial S}{\partial n}\right) = \frac{\varepsilon}{k \ln(N/n - 1)} \qquad (3.8.10)$$

(treating n as a continuous variable). Hence

$$n = N/\{1 + \exp(\varepsilon/kT)\}. \qquad (3.8.11)$$

Substitution in equation (3.8.9) now yields for the internal energy at temperature T

$$U = \frac{N\varepsilon}{1 + \exp(\varepsilon/kT)} + \text{constant} \qquad (3.8.12)$$

and this leads to the result

$$C_V = \partial U/\partial T = Nk\left(\frac{\varepsilon}{kT}\right)^2 \frac{\exp(\varepsilon/kT)}{\{1 + \exp(\varepsilon/kT)\}^2} \qquad (3.8.13)$$

for the heat capacity of the crystal.

More legitimate models for a crystal are described in sections 6.1 and 6.2; however, these lead to equations for C_V which are formally similar to the one we have just reached.

Exercises 3

1. A particle moves along the x-axis with velocity v under no forces. Show that Liouville's equation for the particle is

$$\frac{\partial \rho}{\partial t} + v \frac{\partial \rho}{\partial x} = 0.$$

At $t = 0$, the particle's position is distributed normally about the origin with standard deviation s and its velocity is distributed normally about $v = V$ with standard deviation σ, so that

$$\rho_0 = \frac{1}{2\pi s\sigma} \exp\left[-\frac{x^2}{2s^2} - \frac{(v-V)^2}{2\sigma^2}\right].$$

Prove that

$$\rho = \frac{1}{2\pi s\sigma} \exp\left[-\frac{(x-Vt)^2}{2s^2} - \frac{(v-V)^2}{2\sigma^2} \right].$$

(Hint: The general solution of the Liouville equation is $\phi(x-vt, v)$, where ϕ is an arbitrary function of two variables.)

2. For the particle of the previous exercise, prove that, at time t, the particle's position is distributed normally about $x = Vt$ with standard deviation η given by

$$\eta^2 = s^2 + \sigma^2 t^2.$$

[Note: If the particle's motion is governed by the laws of quantum mechanics and it is represented as a wave packet, then $s\sigma = \hbar/2m$ (Heisenberg's uncertainty principle). It is then of interest that this classical equation for η is valid for quantum mechanics (see e.g. D. F. Lawden, *Mathematical Principles of Quantum Mechanics*, Methuen, 1967, p. 120).]

3. P is the point in phase space representing a specific member of the Gibbs ensemble at time t. D/Dt denotes the rate of change operator as P is followed along its trajectory. Prove that $D\rho/Dt = 0$. (Hint: Differentiate $\rho(q, p, t)$ totally with respect to t and use equations (3.2.1) and (3.3.4).)

4. A closed surface S moves in phase space in such a way that the points lying in its surface always represent the same members of the Gibbs ensemble. Prove that the volume it encloses is conserved. (Hint: Apply the result of Ex. 3 to $\rho \, dV$, where dV is a volume element whose bounding surface has the property of S.)

5. For a certain Maxwell gas, the number of molecules with momenta in the ranges $(p_x, p_x + dp_x)$, etc. is $n(p_x, p_y, p_z) \, dp_x \, dp_y \, dp_z$ per unit volume (the molecules are assumed to be distributed uniformly over the container). Show that the number of molecules with these momenta striking a unit area of the container, with its normal parallel to the negative x-axis, in unit time is $np_x \, dp_x \, dp_y \, dp_z/m$, where m is the molecular mass, provided $p_x > 0$. Deduce that the total force on the element is

$$\frac{1}{m} \int np_x^2 \, dp_x \, dp_y \, dp_z,$$

where the integrations with respect to p_x, p_y, p_z are all over the range $(-\infty, \infty)$. In the case of a gas of point molecules (monatomic gas), deduce that the pressure is given by $P = kNT/V$. [Hint: Use equations (3.7.2), (3.7.4) and (3.7.6).]

6. For a monatomic gas subject to no external field of force, show that the number of molecules with velocities whose magnitudes lie in the range $(c, c + dc)$ and whose directions have spherical polar angles in the ranges $(\theta, \theta + d\theta)$, $(\phi, \phi + d\phi)$, is

$$N\left(\frac{m}{2\pi kT}\right)^{3/2} e^{-mc^2/2kT} c^2 \sin\theta \, dc \, d\theta \, d\phi.$$

Deduce that the number with kinetic energies in the range $(\varepsilon, \varepsilon + d\varepsilon)$ is

$$\frac{2\pi N}{(\pi kT)^{3/2}} e^{-\varepsilon/kT} \varepsilon^{1/2} \, d\varepsilon.$$

7. Using results from the last exercise, show that the mean speed and kinetic energy of a molecule are given by

$$\bar{c} = \sqrt{(8kT/\pi m)}, \qquad \bar{\varepsilon} = 3kT/2,$$

respectively (the last result confirms equation (3.7.10)). If w is the molecular weight for the gas, obtain the formulae

$$\bar{c} = 145.5\sqrt{(T/w)} \, \text{m s}^{-1}; \qquad \bar{\varepsilon} = 2.07T \times 10^{-23} \, \text{J}.$$

8. Show that the most probable value for the speed c of a molecule of a monatomic gas is $\sqrt{(2kT/m)} = 129\sqrt{(T/w)} \, \text{m s}^{-1}$, and its most probable kinetic energy is $\frac{1}{2}kT = 6.9T \times 10^{-24} \, \text{J}$.

CHAPTER 4

Method of mean values

4.1 Statistical analysis of Maxwell–Boltzmann distribution

We shall now examine in more detail the probability distribution (3.5.2) over the range of possible gas conditions satisfying the constraints (3.5.3) and (3.5.4). In particular, we wish to calculate the mean (or expected) values of the numbers of molecules a_1, a_2, etc. in the various cells of the molecular phase space and to find their standard deviations from these means.

It will be helpful to choose an energy unit which is so small that the cell energies ε_i can be taken to be integers. A return to the SI unit (joule) can easily be performed after the required results have been established.

The partition function (3.5.14) is first expressed in the form

$$Z = z^{\varepsilon_1} + z^{\varepsilon_2} + \ldots, \tag{4.1.1}$$

where $z = e^{-\beta}$ and the ε_i are now taken to be integers. The multinomial expansion

$$Z^N = \sum \frac{N!}{a_1! a_2! \ldots} z^{a_1 \varepsilon_1 + a_2 \varepsilon_2 + \ldots}, \tag{4.1.2}$$

where the summation is over all sets $\{a_i\}$ satisfying $\Sigma a_i = N$, then follows by elementary algebra. (Note: If the number of terms in the series for Z is infinite, we must assume the series to be absolutely convergent to validate the necessary rearrangement of terms in the product.) Now, $W = N!/(a_1! a_2! \ldots)$ is the number of members of the Gibbs ensemble modelling the condition of the gas determined by the occupation numbers a_i. Thus, the total number of systems in the ensemble is

$$G = \sum \frac{N!}{a_1! \, a_2! \ldots}, \tag{4.1.3}$$

where the summation is carried out over all sets $\{a_i\}$ satisfying *both* the constraints (3.5.3) and (3.5.4). We note that this sum is the coefficient of z^U in the expansion (4.1.2) of Z^N in powers of z.

Equation (4.1.1) is a Taylor expansion for Z in the neighbourhood of the origin O of the z-plane and therefore defines it as a function of the complex variable z, regular within the circle of convergence Γ of the power series, with centre O. The function Z^N/z^{U+1} has a pole at O and its residue at this pole will be the coefficient of z^U in the expansion of Z^N, i.e. G. It follows that

$$\frac{1}{2\pi i}\oint z^{-U-1}\,Z^N\,dz = G,\qquad(4.1.4)$$

if the contour integral is taken once around O within Γ.

In all cases of interest, both U and N will take very large integral values, although U/N (the average energy per molecule) need not be large. In such circumstances, it is shown in Appendix C that the contour integral (4.1.4) can be very accurately approximated by first identifying the real positive root of the equation (C. 13), viz.

$$g(z) \equiv -\frac{U}{z} + N\frac{Z'}{Z} = 0 \qquad(4.1.5)$$

(primes will denote differentiations with respect to z) and then, with this value of z, equation (4.1.4) can be approximated by

$$\{2\pi g'(z)\}^{-1/2}\, z^{-U-1} Z^N = G. \qquad(4.1.6)$$

Equation (4.1.5) determines z and hence β or the temperature T of the gas. Equation (4.1.6) gives the number of replicas of the system in the Gibbs ensemble.

The mean value of a_1 over all admissible gas conditions is

$$\bar{a}_1 = \sum p a_1, \qquad(4.1.7)$$

where p is the probability of the condition $\{a_i\}$. Substituting for p from equation (3.5.2), we find that

$$\bar{a}_1 = \frac{1}{G}\sum\frac{N!}{(a_1-1)!a_2!\ldots} \qquad(4.1.8)$$

summed over sets $\{a_i\}$ satisfying the two constraints. Putting $a_0 = a_1 - 1$, the constraints can be re-expressed in the form

$$a_0 + a_2 + \ldots = N - 1, \quad a_0\varepsilon_1 + a_2\varepsilon_2 + \ldots = U - \varepsilon_1, \qquad(4.1.9)$$

and then

$$\bar{a}_1 = \frac{N}{G}\sum\frac{(N-1)!}{a_0!a_2!\ldots}, \qquad(4.1.10)$$

the sum being over sets $\{a_0, a_2, \ldots\}$ satisfying the constraints (4.1.9). This sum can be approximated by a contour integral as before to give the result

$$\bar{a}_1 = \frac{N}{G}\{2\pi g'(z)\}^{-1/2}z^{-U+\varepsilon_1-1}Z^{N-1}, \qquad(4.1.11)$$

where, in this case, z satisfies

$$g(z) \equiv -\frac{U-\varepsilon_1}{z} + (N-1)\frac{Z'}{Z} = 0. \qquad(4.1.12)$$

Since ε_1 will be very small by comparison with U, and N will be very large, equation (4.1.12) is virtually the same as equation (4.1.5), and hence z and $g(z)$ may be given the same values in equations (4.1.6) and (4.1.11). Eliminating G between these equations, we obtain

$$\bar{a}_1 = N z^{\varepsilon_1}/Z = N e^{-\beta\varepsilon_1}/Z. \qquad (4.1.13)$$

In general,

$$\bar{a}_i = N e^{-\beta\varepsilon_i}/Z = N e^{-\varepsilon_i/kT}/Z. \qquad (4.1.14)$$

Putting $z = e^{-\beta}$ in equation (4.1.5), it transforms into the earlier equation (3.5.15) determining β. Thus, equation (4.1.14) gives a value for the mean which is identical with the most probable value already calculated (equations (3.5.10), (3.5.16)) and supports the argument of section 3.5.

We next calculate the standard deviation of a_i. Firstly, the mean or expected value of $a_1(a_1 - 1)$ is

$$\mathscr{E}\{a_1(a_1 - 1)\} = \sum p a_1(a_1 - 1) = \frac{N(N-1)}{G} \sum \frac{(N-2)!}{(a_1-2)!\,a_2!\,\ldots}. \qquad (4.1.15)$$

This time we put $a_0 = a_1 - 2$ and write the constraints in the forms

$$a_0 + a_2 + \ldots = N - 2, \qquad a_0\varepsilon_1 + a_2\varepsilon_2 + \ldots = U - 2\varepsilon_1. \qquad (4.1.16)$$

Proceeding as before, we find

$$\mathscr{E}\{a_1(a_1 - 1)\} = \frac{N(N-1)}{G}\{2\pi g'(z)\}^{-1/2} z^{-U+2\varepsilon_1-1} Z^{N-2},$$

$$= N(N-1)z^{2\varepsilon_1}/Z^2, \qquad (4.1.17)$$

having used equation (4.1.6). But,

$$\mathscr{E}\{a_1(a_1 - 1)\} = \mathscr{E}(a_1^2) - \mathscr{E}(a_1). \qquad (4.1.18)$$

Hence, using the result (4.1.13),

$$\mathscr{E}(a_1^2) = \frac{N(N-1)}{Z^2} z^{2\varepsilon_1} + \frac{N}{Z} z^{\varepsilon_1}. \qquad (4.1.19)$$

The variance of a_1 is now given by

$$\mathrm{var}(a_1) = \mathscr{E}(a_1^2) - \bar{a}_1^2 = \frac{N}{Z} z^{\varepsilon_1} - \frac{N}{Z^2} z^{2\varepsilon_1},$$

$$= \bar{a}_1(1 - \bar{a}_1/N) = \bar{a}_1, \qquad (4.1.20)$$

approximately, since a_1/N will be small.

In general,

$$\sigma_i^2 = \mathrm{var}(a_i) = \bar{a}_i, \qquad (4.1.21)$$

where σ_i is the standard deviation of a_i. The relative standard deviation is accordingly

$$\sigma_i/(\bar{a}_i) = 1/\sqrt{(\bar{a}_i)}, \qquad (4.1.22)$$

which is small for cells in which $\bar{a}_i$ is large.

Thus, a cell which can be expected to contain about a million molecules will, on observation, almost certainly be found to contain a number within 0.2 % of this figure; for a cell whose expectation is 16 molecules, however, the corresponding percentage will only be 50 %. But clearly the macroscopic properties of the gas will be dominated by the cells which are densely populated and our analysis therefore indicates that these properties can be accurately calculated by accepting the mean or most probable values of the a_i as actual values, as was done in sections 3.5–3.7.

We conclude this section by noting that, as $T \to +\infty$ and $\beta \to 0$, the mean energy level occupation numbers given by equation (4.1.14) approach equality. If the number g of available energy states is finite (as is possible for quantum molecules in some situations), we can conceive a state in which the number of molecules in each of the energy states is the same, viz. N/g. A state of infinite temperature would there be achieved. Indeed, in these circumstances, negative values of T and β are not ruled out, since Z will remain finite for such values; the normal decline in occupation numbers with increasing energy would then be inverted and the higher energy states would be the more populous. Thus by injecting sufficient energy into such an aggregate of molecules, negative temperature states could be created and these would behave as though their temperatures were *higher* than $+\infty$; e.g. if placed in thermal contact with any body at a positive temperature, heat energy would flow into this body from the body at negative temperature.

4.2 Mixture of gases

Now suppose our isolated container holds two gases in statistical equilibrium with total energy U. A condition of the mixture can be specified by giving the numbers of molecules of the gases whose states are represented by points in the various cells of the two molecular phase spaces. Thus, let a_1, a_2, etc. be the numbers of molecules of the first gas in cells 1, 2, etc. of its phase space and let b_1, b_2, etc. be the numbers of molecules of the second gas in cells 1, 2, etc. of the other phase space. Let ε_1, ε_2, etc. be the molecular energies associated with the first set of cells and η_1, η_2, etc. the molecular energies corresponding to the second set. If M is the number of molecules for the first gas and N is the number for the second, the a_i, b_j must satisfy the constraints

$$\sum a_i = M, \qquad \sum b_j = N, \qquad \sum a_i \varepsilon_i + \sum b_j \eta_j = U. \qquad (4.2.1)$$

The number of members of the microcanonical ensemble modelling this condition will be

$$\frac{M!}{a_1! a_2! \ldots} \cdot \frac{N!}{b_1! b_2! \ldots} \qquad (4.2.2)$$

and the probability for the condition is accordingly given by

$$p = \frac{M!}{a_1! a_2! \ldots} \cdot \frac{N!}{b_1! b_2! \ldots} \cdot \frac{1}{G} \qquad (4.2.3)$$

where G is the total number of members of the ensemble.

Introducing partition functions

$$Z_1 = \sum z^{\varepsilon_i}, \qquad Z_2 = \sum z^{\eta_j}, \qquad (4.2.4)$$

where $z = e^{-\beta}$, for the two types of molecule, for sufficiently small $|z|$ we have the expansion

$$Z_1^M Z_2^N = \sum \frac{M!}{a_1! a_2! \ldots} \cdot \frac{N!}{b_1! b_2! \ldots} z^{a_1 \varepsilon_1 + a_2 \varepsilon_2 + \ldots + b_1 \eta_1 + b_2 \eta_2 + \ldots}, \qquad (4.2.5)$$

the summation being over all sets $\{a_i\}, \{b_j\}$ satisfying the constraints $\Sigma a_i = M$, $\Sigma b_j = N$. It follows, as in section 4.1, that the coefficient of z^U in this expansion is the sum

$$\sum \frac{M!}{a_1! a_2! \ldots} \cdot \frac{N!}{b_1! b_2! \ldots} \qquad (4.2.6)$$

extended over all sets $\{a_i\}, \{b_j\}$ satisfying the constraints (4.2.1). Since $\Sigma p = 1$, equation (4.2.3) shows that this coefficient must be G.

It now follows that

$$\frac{1}{2\pi\iota} \oint z^{-U-1} Z_1^M Z_2^N \, dz = G \qquad (4.2.7)$$

(cf. equation (4.1.4)), the contour integral being taken around a contour encircling $z = 0$ once. Using the result found in Appendix C again, as in section 4.1, we conclude that

$$\{2\pi g'(z)\}^{-1/2} z^{-U-1} Z_1^M Z_2^N = G, \qquad (4.2.8)$$

where

$$g(z) \equiv -\frac{U}{z} + M\frac{Z_1'}{Z_1} + N\frac{Z_2'}{Z_2} = 0 \qquad (4.2.9)$$

determines z and $g(z)$. Putting $z = e^{-\beta}$, this last equation can be expressed in the form

$$\frac{\partial}{\partial \beta}(M \ln Z_1 + N \ln Z_2) = -U. \qquad (4.2.10)$$

(Cf. equation (3.5.15).)

The mean value of a_1 over the probability distribution (4.2.3) is given by

$$\bar{a}_1 = \sum p a_1 = \frac{1}{G} \sum \frac{M!}{(a_1 - 1)! a_2! \ldots} \cdot \frac{N!}{b_1! b_2! \ldots},$$

$$= \frac{M}{G} \sum \frac{(M-1)!}{a_0! a_2! \ldots} \cdot \frac{N!}{b_1! b_2! \ldots}, \qquad (4.2.11)$$

where $a_0 = a_1 - 1$, the last sum being taken over sets $\{a_0, a_2, \ldots\}$, $\{b_1, b_2, \ldots\}$ satisfying the constraints

$$\left. \begin{array}{l} a_0 + a_2 + \ldots = M - 1, \quad b_1 + b_2 + \ldots = N, \\ a_0\varepsilon_1 + a_2\varepsilon_2 + \ldots + b_1\eta_1 + b_2\eta_2 + \ldots = U - \varepsilon_1. \end{array} \right\} \quad (4.2.12)$$

Thus, by an argument similar to that used to derive equation (4.1.11),

$$\bar{a}_1 = \frac{M}{G} \oint z^{-U+\varepsilon_1-1} Z_1^{M-1} Z_2^N \, dz,$$

$$= \frac{M}{G} \{2\pi g'(z)\}^{-1/2} z^{-U+\varepsilon_1-1} Z_1^{M-1} Z_2^N, \quad (4.2.13)$$

where z, $g(z)$ are again determined by equation (4.2.9). Equations (4.2.8), (4.2.13) now lead to the formula

$$\bar{a}_1 = \frac{M}{Z_1} z^{\varepsilon_1} = \frac{M}{Z_1} e^{-\beta\varepsilon_1}. \quad (4.2.14)$$

Similar calculations lead to the general results

$$\bar{a}_i = \frac{M}{Z_1} e^{-\beta\varepsilon_i}, \quad \bar{b}_j = \frac{N}{Z_2} e^{-\beta\eta_j}. \quad (4.2.15)$$

The argument of section 3.6 can now be repeated to derive equations (3.6.6) once again, with

$$\Psi = kM \ln Z_1 + kN \ln Z_2 \quad (4.2.16)$$

replacing equation (3.6.7). β is now determined by equation (4.2.10) and, as before, $\beta = 1/kT$ gives the absolute temperature of the mixture.

The pressure of the mixture follows from equation (3.6.13) thus:

$$P = T\partial\Psi/\partial V = kMT\frac{\partial}{\partial V}(\ln Z_1) + kNT\frac{\partial}{\partial V}(\ln Z_2). \quad (4.2.17)$$

Clearly, the partial pressure of each gas is equal to the pressure it would exert if it alone occupied the container at temperature T; this establishes *Dalton's law*.

The form taken by equation (4.2.16) indicates that we can, in all circumstances, treat the two gases as occupying the container independently of one another; this is, of course, a consequence of our assumption that the interaction between gas molecules is negligible. Thus, if the molecules of both gases can be treated as point particles (monatomic molecules), Maxwell's distribution law for molecular velocities can be established for each gas separately, as in section 3.7 and the partition functions Z_1, Z_2 each calculated by a formula of the type (3.7.9). Equations for the internal energy and state of the mixture then follow in the forms

$$U = \tfrac{3}{2}(M+N)kT, \quad PV = k(M+N)T. \quad (4.2.18)$$

The extension of all these results to the case of a mixture of any number of gases will be obvious.

4.3 System in a heat bath. Canonical ensemble

We are now ready to study the case of a system placed in an environment at constant temperature T and which interacts with its surroundings to reach a state of statistical equilibrium at the same temperature. Thus this system is no longer isolated and prohibited from engaging in energy transactions across the walls of its container; its circumstances are accordingly much closer to those which prevail for systems normally met with in practice. It will be convenient to model the system's environment by an atmosphere comprising a mixture of gases at uniform temperature T; our calculations will assume only two gases are involved, but the extension to a larger number is trivial. Of course, we know from experience that the nature of the heat bath can have no influence on the state of our system, which will depend on the temperature of the environment alone. This, as expected, will prove to be the case.

Evidently, there can be no objection to assuming the atmosphere is isolated from the rest of the universe; this can have no effect on the system we are studying. Thus, we have an isolated system I, which comprises (i) a mixture of two gases together with (ii) the system S which is of primary interest, all in statistical equilibrium. Possible conditions of the two gases will be specified using the nomenclature of the previous section. The system S can be of any type and is not necessarily a gas or even separable into molecules or other subsystems; neither is it necessarily a macroscopic system but may, in particular cases, be a single atom or molecule. Its phase space will be divided into small cells in the usual way and the energy of S when in the kth cell will be denoted by e_k (not, necessarily, all different). n_k will denote the occupation number for S in the kth cell; thus for a state of S which puts its representative point into cell 1, we shall have $n_1 = 1$ and $n_k = 0$, $k \neq 1$ (i.e. for all states of S, all the n_k are zero except one, which has the value 1). The total energy of I will be assumed known, and will be denoted by E. Every condition of I must then satisfy the constraints

$$\sum a_i = M, \quad \sum b_j = N, \quad \sum n_k = 1,$$
$$\sum a_i \varepsilon_i + \sum b_j \eta_j + \sum n_k e_k = E. \tag{4.3.1}$$

As usual, we adopt an energy unit so small that ε_i, η_j, e_k are integers.

Any condition $\{a_i, b_j, n_k\}$ of I is modelled by

$$\frac{M!}{a_1! a_2! \ldots} \cdot \frac{N!}{b_1! b_2! \ldots} \tag{4.3.2}$$

members of a microcanonical ensemble. Thus, if G is the number of members of this ensemble (all equally probable, remember) satisfying the constraints (4.3.1), the probability of the condition $\{a_i, b_j, n_k\}$ is

$$p = \frac{1}{G} \cdot \frac{M!}{a_1! a_2! \ldots} \cdot \frac{N!}{b_1! b_2! \ldots}. \tag{4.3.3}$$

Also

$$G = \sum \frac{M!}{a_1! a_2! \ldots} \cdot \frac{N!}{b_1! b_2! \ldots}, \tag{4.3.4}$$

the summation being over sets of integers $\{a_i, b_j, n_k\}$ satisfying the constraints (4.3.1).

We next introduce, as in the previous section, partition functions Z_1, Z_2 for the gas molecules and a further function for S, viz.

$$Z = \sum z^{e_k}. \tag{4.3.5}$$

The analysis now follows along the lines of section 4.2. Thus, consider the expansion

$$Z_1^M Z_2^N Z = \sum \frac{M!}{a_1! a_2! \ldots} \cdot \frac{N!}{b_1! b_2! \ldots} z^{\Sigma a_i \varepsilon_i + \Sigma b_i \eta_j + \Sigma n_k e_k}, \tag{4.3.6}$$

the summation being over sets $\{a_i, b_j, n_k\}$ satisfying $\Sigma a_i = M, \Sigma b_j = N, \Sigma n_k = 1$. It follows from equation (4.3.4) that

$$G = \text{coefficient of } z^E \text{ in the expansion of } Z_1^M Z_2^N Z,$$

$$= \frac{1}{2\pi i} \oint z^{-E-1} Z_1^M Z_2^N Z \, dz,$$

$$= \{2\pi g'(z)\}^{-1/2} z^{-E-1} Z_1^M Z_2^N Z, \tag{4.3.7}$$

where $z, g(z)$ are given by

$$g(z) \equiv -\frac{E}{z} + M \frac{Z_1'}{Z_1} + N \frac{Z_2'}{Z_2} = 0 \tag{4.3.8}$$

for large E, M and N. Putting $z = e^{-\beta}$, this equation becomes

$$\frac{\partial}{\partial \beta} (M \ln Z_1 + N \ln Z_2) = -E \tag{4.3.9}$$

determining β and hence the temperature T of the system. Note that Z is absent from this equation; this reflects our assumption that M, N are very large, so that the atmosphere (or heat bath) is dominant in determining the temperature.

For members of the ensemble, n_k will take values 1 or 0, depending on whether the state of S lies in the kth cell of its phase space or not. Thus, the mean $\bar{n}_k$ over the ensemble will give the proportion of its members for which S's state lies in the kth cell, i.e. the probability of this event. Now

$$\bar{n}_k = \sum p n_k = \frac{1}{G} \sum \frac{M!}{a_1! a_2! \ldots} \cdot \frac{N!}{b_1! b_2! \ldots} \cdot n_k \tag{4.3.10}$$

summed over sets $\{a_i, b_j, n_k\}$ satisfying the constraints (4.3.1). Omitting terms from the summation in which $n_k = 0$, it will be seen that

$$\bar{n}_k = \frac{1}{G} \sum \frac{M!}{a_1! a_2! \ldots} \cdot \frac{N!}{b_1! b_2! \ldots}, \tag{4.3.11}$$

summed over sets $\{a_i, b_j\}$ satisfying

$$\sum a_i = M, \qquad \sum b_j = N, \qquad \sum a_i \varepsilon_i + \sum b_j \eta_j = E - e_k. \tag{4.3.12}$$

Hence

$$\bar{n}_k = \frac{1}{G} \times \text{coefficient of } z^{E-e_k} \text{ in the expansion of } Z_1^M Z_2^N,$$

$$= \frac{1}{G} \cdot \frac{1}{2\pi i} \oint z^{-E+e_k-1} Z_1^M Z_2^N \, dz,$$

$$= \frac{1}{G} \{2\pi g'(z)\}^{-1/2} z^{-E+e_k-1} Z_1^M Z_2^N, \qquad (4.3.13)$$

where z, $g(z)$ are now determined by the equations

$$g(z) \equiv -\frac{E-e_k}{z} + M\frac{Z_1'}{Z_1} + N\frac{Z_2'}{Z_2} = 0. \qquad (4.3.14)$$

Since we can assume e_k to be small by comparison with E, there is negligible error in continuing to regard z, $g(z)$ as being fixed by equations (4.3.8). Equations (4.3.7), (4.3.13) then lead to the result.

$$\bar{n}_k = z^{e_k}/Z = e^{-\beta e_k}/Z = -\frac{1}{\beta}\frac{\partial}{\partial e_k}(\ln Z) \qquad (4.3.15)$$

for the probability a system S, placed in a heat bath at temperature T, will be found in a state with energy e_k, corresponding to the kth cell of its phase space.

Note that the nature of the heat bath does not enter into this formula, which is very satisfactory and in accord with expectations. This means that, in such circumstances, we can ignore the environment and model S by a Gibbs ensemble comprising a very large number of replicas of S covering all possible states, the number of these in the kth state being chosen to be proportional to $\exp(-e_k/kT)$. This is termed the *canonical ensemble* and replaces the earlier microcanonical ensemble for the generality of systems whose temperature, rather than total energy, is given.

For such a system, the total energy e $(= \Sigma n_k e_k)$ is not known precisely since, not being isolated from the environment, this can act as a heat source or sink. However, the mean energy U is easily calculated as follows:

$$U = \bar{e} = \sum \bar{n}_k e_k = \frac{1}{Z}\sum e_k \exp(-\beta e_k) = -\frac{1}{Z}\frac{\partial Z}{\partial \beta} = -\frac{\partial}{\partial \beta}(\ln Z) \quad (4.3.16)$$

The second moment of the energy is

$$\sum \bar{n}_k e_k^2 = \frac{1}{Z}\sum e_k^2 \exp(-\beta e_k) = \frac{1}{Z}\frac{\partial^2 Z}{\partial \beta^2} \qquad (4.3.17)$$

and thus the variance is

$$\text{var}(e) = \frac{1}{Z}\frac{\partial^2 Z}{\partial \beta^2} - \bar{e}^2 = \frac{1}{Z}\frac{\partial^2 Z}{\partial \beta^2} - \frac{1}{Z^2}\left(\frac{\partial Z}{\partial \beta}\right)^2,$$

$$= \frac{\partial^2}{\partial \beta^2}(\ln Z). \qquad (4.3.18)$$

This result can also be expressed in the form

$$\text{var}\,(e) = -\frac{\partial U}{\partial \beta} = kT^2 \frac{\partial U}{\partial T} = kT^2 C_V, \tag{4.3.19}$$

where $C_V = \partial U/\partial T$ is the heat capacity at constant volume for S (see sections 1.5 and 2.3). (Note: In the last equation, the differentiation with respect to T is performed keeping all the cell energies e_k constant; this means that all the external fields of force to which S is subject are not permitted to vary and, in particular, the walls of the container (if any) are not moved, i.e. V is constant.)

It is shown in the next section that, for a gas comprising N monatomic molecules whose interaction energy is negligible, $U = 3NkT/2$. For such a system, therefore, var $(e) = 3Nk^2T^2/2$ and the relative standard deviation for the energy is

$$\sqrt{(\text{var}\,e)}/\bar{e} = \sqrt{(2/3N)}, \tag{4.3.20}$$

which is very small when N is large. Thus, the energy distribution is very sharp, as it is for most systems with a very large number of degrees of freedom. In these circumstances, it is permissible to regard the energy as taking the value U precisely and the distinction between given-energy and given-temperature systems largely disappears.

As a particular case, any one of the N molecules comprising the gas considered in section 3.5 can be looked upon as the system S immersed in the heat bath provided by the other $(N-1)$ molecules. The partition function for this molecule has been given at equation (3.5.14) and the probability of finding it in the energy state ε_i is therefore given by equation (4.3.15) to be $e^{-\beta \varepsilon_i}/Z$. Thus, we expect to find $Ne^{-\beta \varepsilon_i}/Z$ molecules of the gas in this state and this is precisely the result (3.5.10).

The argument of section 3.6 following equation (3.6.1) can be applied to mean values over the canonical ensemble, thus:

$$dU = d(\sum \bar{n}_k e_k) = \sum \bar{n}_k de_k + \sum e_k \, d\bar{n}_k; \tag{4.3.21}$$

the first term on the right gives the mean work done by the external forces on the system and the second term represents the mean heat energy $d\bar{Q}$ supplied. Hence

$$\beta \, d\bar{Q} = \beta \, dU - \beta \sum \bar{n}_k de_k,$$

$$= d(\beta U) - U \, d\beta + \sum \frac{\partial}{\partial e_k} (\ln Z) \, de_k$$

$$= d(\beta U + \ln Z), \tag{4.3.22}$$

having used equations (4.3.15) and (4.3.16). This shows that $\beta \, d\bar{Q}$ is an exact differential and, assuming we can identify mean values with the macroscopic state variables of thermodynamics, we conclude that

$$\beta = 1/kT, \quad S = k \ln Z + U/T, \tag{4.3.23}$$

where S is the entropy of the system.

The first term in the right-hand member of equation (4.3.21) is the work done on S by the external fields. Thus, if θ_r are generalized coordinates determining the configuration of the external fields, and $\overline{F}_r$ are the corresponding mean generalized force components applied by the system to these fields, the average work done by the system for increments $d\theta_r$ in the coordinates is

$$\sum_r \overline{F}_r \, d\theta_r = -\sum_k \bar{n}_k \, de_k = \frac{1}{\beta} \sum_r \frac{\partial}{\partial \theta_r} (\ln Z) \, d\theta_r \qquad (4.3.24)$$

by use of equation (4.3.15) (cf. equation (3.6.11)). It follows that

$$\overline{F}_r = \frac{1}{\beta} \frac{\partial}{\partial \theta_r} (\ln Z). \qquad (4.3.25)$$

Equations (4.3.15), (4.3.16), (4.3.23) and (4.3.25) demonstrate that a knowledge of the partition function Z is all that is necessary for a determination of the macroscopic state variables for the system and the calculation of Z is accordingly a seminal problem for statistical mechanics.

4.4 Partition functions for mixtures

Suppose we have a pair of non-interacting systems at the same temperature, such as a dilute mixture of two gases. Let ε_i $(i = 1, 2, \ldots)$ be the energy states for the first system and η_j $(j = 1, 2, \ldots)$ those for the second system, leading to partition functions

$$Z_1 = \sum e^{-\beta \varepsilon_i}, \quad Z_2 = \sum e^{-\beta \eta_j}. \qquad (4.4.1)$$

There is clearly no objection to considering the systems together as forming a single system whose energy states are $\{\varepsilon_i + \eta_j\}$ and whose partition function is therefore

$$\sum_{i,j} e^{-\beta(\varepsilon_i + \eta_j)} = \sum_i e^{-\beta \varepsilon_i} \times \sum_j e^{-\beta \eta_j} = Z_1 Z_2. \qquad (4.4.2)$$

We conclude that, when non-interacting systems at the same temperature are combined to form a composite system, their partition functions are multiplied.

For example, if Z (equation (3.5.14)) is the partition function for a molecule of the gas studied in section 3.5, the partition function for the whole gas of N non-interacting molecules is Z^N. If the gas is kept at temperature T, equation (4.3.16) shows that its mean energy is

$$U = -\frac{\partial}{\partial \beta} (\ln Z^N) = -N \frac{\partial}{\partial \beta} (\ln Z), \qquad (4.4.3)$$

i.e. N times the mean energy per molecule. It was shown in section 3.7 (equation (3.7.9)) that, for a gas of point molecules, $Z = \alpha T^{3/2}$, where α is independent of T. It follows that

$$U = NkT^2 \frac{\partial}{\partial T} (\ln Z) = \tfrac{3}{2} NkT, \qquad (4.4.4)$$

i.e. a mean energy of $3kT/2$ per molecule.

As another example, consider a mixture of two gases, one with M molecules each having partition function Z_1 and the other having N molecules each with partition function Z_2. Then the partition function for the mixture is $Z_1^M Z_2^N$ and the equation (4.3.16) gives for the overall mean energy

$$U = -\frac{\partial}{\partial \beta}(M \ln Z_1 + N \ln Z_2). \tag{4.4.5}$$

This equation has precisely the same form as the equation (4.2.10) determining the temperature of the mixture when its energy U is precisely determined. This confirms that it is permissible to regard both the temperature and energy of a system placed in a heat bath as given quantities. Note also that the last equation implies

$$U = U_1 + U_2, \tag{4.4.6}$$

where U_1, U_2 are the mean energies of the respective gas components.

Consider a system comprising a very large number N of similar subsystems, each with a partition function Z. Assuming the interaction between subsystems to be negligible, the partition function for the composite system is Z^N. The mean and variance of the system's energy are given by equations (4.3.16), (4.3.18) to be

$$U = -N\frac{\partial}{\partial \beta}(\ln Z), \quad \text{var}(e) = N\frac{\partial^2}{\partial \beta^2}(\ln Z) \tag{4.4.7}$$

and the relative standard deviation for e is accordingly

$$\sqrt{\{\text{var}(e)\}}/U \propto 1/\sqrt{N}. \tag{4.4.8}$$

This tends to zero as $N \to \infty$, showing once again that the energy distribution for such systems is sharply peaked.

Such a system is the model for a crystal already analysed in section 3.8. Each atom has two energy states E and $E + \varepsilon$, and its partition function is accordingly

$$Z = e^{-\beta E} + e^{-\beta(E+\varepsilon)} = e^{-\beta E}(1 + e^{-\beta \varepsilon}). \tag{4.4.9}$$

The mean energy of the crystal is now given by the first of equations (4.4.7) to be

$$U = NE - N\frac{\partial}{\partial \beta}\ln(1 + e^{-\beta \varepsilon}) = NE + \frac{N\varepsilon}{1 + \exp(\beta \varepsilon)}. \tag{4.4.10}$$

This is the same result as previously found at equations (3.8.9) and (3.8.11), except that the earlier calculation regarded the crystal as isolated and therefore interpreted U as the sharp energy of the crystal.

It sometimes happens that the energy e of a system breaks up into the sum of a number of parts, each of which is a function of phase space variables not present in the others. For example, the energy of a rigid molecule separates into (i) the P.E. of the molecule in the external field, (ii) the K.E. of its translatory motion and (iii) the K.E. of its rotational motion; if (x, y, z) are the coordinates of its mass centre, (p_x, p_y, p_z) are the components of its linear momentum, (θ, ϕ, ψ) are Euler angles for the rotational motion about the mass centre and $(p_\theta, p_\phi, p_\psi)$ are the associated components of momentum, then energy (i) is a function of (x, y, z), energy (ii) is a function of (p_x, p_y, p_z) and energy (iii) is a function of $(\theta, \phi, \psi, p_\theta, p_\phi, p_\psi)$. In these circumstances, the phase space variables separate into groups, each of which defines a phase subspace which can be divided into cells in the usual way. Then the state of the system can be specified by naming the cell in each subspace occupied by the point representing the values of the corresponding group of variables. Thus, if $e = \varepsilon + \eta + \ldots$ and ε_i is the value of ε for the ith cell of the first subspace, η_j is the value of η for the jth cell of the second subspace, and so on, then the partition function for the system is

$$Z = \sum_{i, j, \ldots} z^{\varepsilon_i + \eta_j + \cdots} = \left(\sum_i z^{\varepsilon_i}\right)\left(\sum_j z^{\eta_j}\right) \ldots = Z_1 Z_2 \ldots, \qquad (4.4.11)$$

i.e. the partition function breaks up into factors $Z_1, Z_2, \ldots$, which are individual partition functions for the different components of e. An example follows in the next section.

4.5 Gas of diatomic molecules

Suppose we replace the gas of point molecules already analysed in section 3.7 by one whose N molecules can each be modelled as a pair of point atoms, rigidly held together at a constant distance apart. The molecule will be assumed to have zero moment of inertia about the line of its atoms and moment of inertia C about any axis through the mass centre G perpendicular to this line. The kinetic energy of spin about G can no longer be neglected when calculating the partition function. However, we shall still assume the gas to be sufficiently dilute, so that the energy of molecular interaction can be ignored. For simplicity, we shall disregard any external field of force.

Taking rectangular Cartesian axes $Oxyz$, let $GXYZ$ be parallel axes through the mass centre of the molecule (Fig. 4.1). If GL is the axis of the molecule, suppose the plane GLZ intersects the plane GXY in the line GP. Then we can specify the orientation of the molecule by the angles $LGZ = \theta$ and $PGX = \phi$.

Let GM be perpendicular to GL and lie in the plane LGZ and let GN be perpendicular to GL and lie in the plane GXY. Then GL, GM, GN are principal axes of inertia for the molecule, with principal moments of inertia $(0, C, C)$. The molecule has angular velocities $\dot{\theta}$ about GN and $\dot{\phi}$ about GZ. Resolving these along the principal axes, we get components $(\dot{\phi} \cos \theta, -\dot{\phi} \sin \theta, \dot{\theta})$ and the K.E. of the rotational motion about G is accordingly

$$L = \tfrac{1}{2}C(\dot{\phi}^2 \sin^2\theta + \dot{\theta}^2). \qquad (4.5.1)$$

Since no external forces act upon the molecule, this is also the Lagrangian for the rotational motion. The momentum components associated with the coordinates (θ, ϕ) are now calculable thus:

$$p_\theta = \frac{\partial L}{\partial \dot{\theta}} = C\dot{\theta}, \quad p_\phi = \frac{\partial L}{\partial \dot{\phi}} = C\dot{\phi} \sin^2 \theta. \qquad (4.5.2)$$

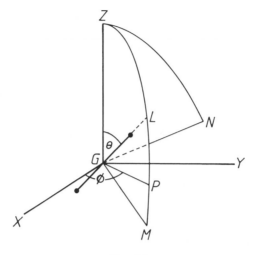

Fig. 4.1

Eliminating θ, $\dot{\phi}$ between equations (4.5.1), (4.5.2), we find the Hamiltonian for the rotational motion to be

$$H = L = \frac{1}{2C}(p_\theta^2 + p_\phi^2 \operatorname{cosec}^2 \theta). \tag{4.5.3}$$

The energy for the whole motion of the molecule can now be expressed as

$$e = \frac{1}{2m}(p_x^2 + p_y^2 + p_z^2) + \frac{1}{2C}(p_\theta^2 + p_\phi^2 \operatorname{cosec}^2 \theta) \tag{4.5.4}$$

and, by the principle explained in the previous section, we see that the molecule's partition function separates into two factors, (i) for the motion of G, and (ii) for the rotational motion about G. The factor for the first component has already been calculated at equation (3.7.9) and $Z_1 = h^{-3}V(2\pi mkT)^{3/2}$. The factor for the second component is

$$Z_2 = h^{-2}\int \exp(-\beta H)\,d\theta\,d\phi\,dp_\theta\,dp_\phi, \tag{4.5.5}$$

where H is given by equation (4.5.3) and the limits for the respective variables of integration are $(0, \pi)$, $(0, 2\pi)$, $(-\infty, \infty)$, $(-\infty, \infty)$.

Integration with respect to ϕ is trivial and yields a factor 2π. Integration with respect to p_θ is next performed by using the identity (3.7.4); the result is

$$Z_2 = 2\pi h^{-2}(2\pi C/\beta)^{1/2}\int \exp(-\beta p_\phi^2 \operatorname{cosec}^2\theta/2C)\,d\theta\,dp_\phi. \tag{4.5.6}$$

Integration with respect to p_ϕ follows from the same identity, to give

$$Z_2 = 2\pi h^{-2}(2\pi C/\beta)\int_0^\pi \sin\theta\,d\theta = 8\pi^2 h^{-2}CkT. \tag{4.5.7}$$

The complete partition function can now be written down as

$$Z = Z_1 Z_2 = 16\pi^3 \sqrt{(2\pi)}h^{-5} VCm^{3/2} (kT)^{5/2}. \qquad (4.5.8)$$

Since Z^N is the partition function for the whole gas, equation (4.3.16) now gives for the mean energy at temperature T,

$$U = 5NkT/2. \qquad (4.5.9)$$

Comparing this result with equation (4.4.4) for a monatomic gas, we see that the energy per molecule has been increased by kT due to the rotational component. The pressure can be calculated from equation (3.6.13) and this leads to the ideal gas law (1.5.4), i.e. the rotational motion has no effect on the pressure.

4.6 Paramagnetic materials. Langevin's function

The molecules of a paramagnetic material are magnetic dipoles which, in the absence of an external field, have their axes oriented randomly, thus yielding no overall resultant magnetic moment over any volume of the substance having macroscopic dimensions. When placed in a magnetic field, the dipoles tend to align themselves with the field and the material becomes magnetized. This tendency is, however, opposed by the random motions of the molecules associated with heat energy and we expect that, for a given strength of the applied field, the intensity of magnetization will be reduced as the temperature rises—this expectation is supported by Curie's law (see section 2.1).

Let **m** be the moment of a molecular dipole and **H** the intensity of the applied field (assumed uniform). Then, the potential energy of the dipole in the field (using SI units) is $-\mu_0 \mathbf{m} \cdot \mathbf{H} = -\mu_0 mH \cos\theta$, where θ is the angle made by the dipole axis with the field. We shall assume that the molecule behaves mechanically like a rigid straight rod, with principal moments of inertia $(0, C, C)$ as in the case of the diatomic gas molecule studied in the previous section. We shall also assume that the molecule's mass centre is securely anchored, but that it is free to pivot about its mass centre. Interaction between the molecules will be ignored. The energy of such interactions and of the vibrations of the mass centres can be allowed for (see sections 6.1 and 6.2) separately and contributes to the heat capacity of the material; however, its contribution to the material's magnetic properties is minimal. Then, defining the angles θ and ϕ as in the last section, the energy for the molecule is given by

$$e = -\mu_0 mH \cos\theta + \frac{1}{2C}(p_\theta^2 + p_\phi^2 \csc^2\theta), \qquad (4.6.1)$$

where the external field has been taken in the direction $\theta = 0$.

If the material is in equilibrium at the temperature T, we can regard any one of the molecules to be a system in equilibrium in the heat bath provided by all the other molecules. Then, by equation (4.3.15), the probability of finding the point representing the molecule's state in the phase space cell $(\theta, \theta + d\theta)$, $(\phi, \phi + d\phi)$, $(p_\theta, p_\theta + dp_\theta)$, $(p_\phi, p_\phi + dp_\phi)$, is proportional to

$$\exp\{\beta\mu_0 mH \cos\theta - \beta(p_\theta^2 + p_\phi^2 \csc^2\theta)/2C\} \cdot d\theta \, d\phi \, dp_\theta \, dp_\phi, \qquad (4.6.2)$$

where, as usual, $\beta = 1/kT$. Integrating this expression with respect to ϕ over the range $(0, 2\pi)$, and with respect to p_θ and p_ϕ over the range $(-\infty, \infty)$ (using the identity (3.7.4)), we calculate that the probability of finding θ in the interval $(\theta, \theta + d\theta)$ is

$$A \exp\left(\beta\mu_0\, mH \cos\theta\right)\cdot\sin\theta\, d\theta, \tag{4.6.3}$$

where A is a factor of proportionality independent of θ. Since the total probability of finding θ in the interval $(0, \pi)$ must be unity, by integrating the last expression over this range, we find that

$$1 = A \int_0^\pi \exp\left(\beta\mu_0\, mH \cos\theta\right)\cdot\sin\theta\, d\theta$$

$$= A \int_{-1}^1 \exp\left(\beta\mu_0\, mHx\right) dx = \frac{2A}{\alpha}\sinh\alpha, \tag{4.6.4}$$

where $x = \cos\theta$ and

$$\alpha = \beta\mu_0\, mH = \mu_0\, mH/kT. \tag{4.6.5}$$

Thus, the probability θ lies in $(\theta, \theta + d\theta)$ is

$$dP = \tfrac{1}{2}\alpha \operatorname{cosech}\alpha \cdot \exp\left(\alpha\cos\theta\right)\cdot\sin\theta\, d\theta. \tag{4.6.6}$$

Clearly, only the component of the molecule's dipole moment in the direction of the field will contribute to the material's overall magnetization—the other components will have zero means. The mean of this component is now found to be

$$\int_0^1 m\cos\theta\, dP = \tfrac{1}{2}m\alpha \operatorname{cosech}\alpha \int_0^\pi e^{\alpha\cos\theta}\sin\theta\cos\theta\, d\theta$$

$$= \tfrac{1}{2}m\alpha \operatorname{cosech}\alpha \int_{-1}^1 xe^{\alpha x}\, dx$$

$$= m\left(\coth\alpha - 1/\alpha\right). \tag{4.6.7}$$

Hence, if the paramagnetic specimen contains N molecules per unit volume, the observed intensity of magnetization will be given by

$$I = Nm\left(\coth\alpha - 1/\alpha\right). \tag{4.6.8}$$

If α is large (i.e. intense applied field or low temperature), then $\coth\alpha = 1$, $1/\alpha = 0$ and $I = I_\infty = Nm$; clearly, almost all the dipoles are aligned with the field and the material has reached magnetic saturation. We can now write

$$I/I_\infty = \coth\alpha - 1/\alpha. \tag{4.6.9}$$

This function of α is called *Langevin's function* (P. Langevin, 1872–1946); it is graphed in Fig. 4.2. For small α (i.e. weak field or high temperature) it can be approximated by $\alpha/3$ and thus

$$I = \tfrac{1}{3}\alpha I_\infty = \mu_0\, Nm^2 H/(3kT). \tag{4.6.10}$$

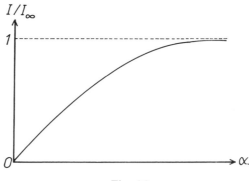

Fig. 4.2

This shows that, in these circumstances $I = \chi H$, where

$$\chi = \mu_0 \, Nm^2/(3kT) \qquad (4.6.11)$$

is the magnetic susceptibility. Since $\chi \propto 1/T$, we have established Curie's law for paramagnetic materials which are far from magnetic saturation.

If, instead of the magnetization, we calculate M the moment per mole, then $N = L$ (Avogadro's number), $M_\infty = Lm$ and

$$\chi = \mu_0 M_\infty^2/(3RT), \qquad (4.6.12)$$

where $R = kL$ is the gas constant (section 1.5).

4.7 Additional constraints

Thus far, it has been assumed that the only information we have in regard to the state of a system is either its internal energy or its temperature. Thus in section 3.5 the total energy U of a gas was prescribed and this datum, together with a supposed knowledge of the total number of molecules N, were represented by the two constraints (3.5.3), (3.5.4). However, further information relating to the macroscopic state of a system is usually available and this should be taken into account by subjecting the mathematical analysis to further appropriate constraints.

As an example, the gas of section 3.5 will normally be supposed isolated in a stationary container and, when statistical equilibrium has been attained, the gas's overall linear momentum will accordingly be zero. Hence, if p_{ix} is the x-component of momentum in the ith cell of molecular phase space, the constraint expressing the vanishing of the sum over all the molecules of these components for a gas in the condition $\{a_i\}$ is

$$\sum a_i p_{ix} = 0. \qquad (4.7.1)$$

Two similar constraints can be written down expressing the vanishing of the y- and z-components of the resultant momentum. Thus, the maximization of the probability p (equation (3.5.2)) should have been subjected to these constraints in addition to the constraints (3.5.3), (3.5.4), which were actually taken into account. The reason why the additional constraints could be safely ignored was that, luckily, the result obtained by so doing is found to satisfy the requirement of vanishing overall momentum. For, it is a consequence of equation (3.7.2) that the number of molecules with momentum components (p_x, p_y, p_z) is the same as the number with components $(-p_x, p_y, p_z)$ and the x-components of momentum therefore cancel. Similarly for the other two components. Clearly, therefore, if the Maxwell-distribution maximizes p when the additional constraints are ignored, it must also do so under the more restrictive conditions which prevail when they are taken into account.

Further, the number of molecules in a neighbourhood of (x, y, z) with momentum components (p_x, p_y, p_z) is also the same as the number in this neighbourhood with components $(-p_x, -p_y, p_z)$ and their combined angular momentum $\Sigma (x p_y - y p_x)$ about any axis parallel to the z-axis will therefore be zero. It follows that, the constraint that the overall angular momentum of the gas about any such axis should be zero, is also automatically satisfied. This justifies our ignoring a further piece of information which is usually available in this case.

If, therefore, for a given system, the energy or temperature constraint can reasonably be augmented by others, we may prefer to ignore these for the time being and to check later whether they are fortuitously satisfied by the result of our analysis. If this proves to be the case, our earlier disregard of the additional constraints is thereby justified and the result is acceptable. If, however, these constraints are violated, it will be necessary to recalculate the problem by a method which takes account of all the constraints from the outset. Such a method will now be described.

Consider, once again, the gas of N molecules as described in section 3.5 and suppose we require an admissible condition of the gas to satisfy the constraints (3.5.3), (3.5.4) and a further constraint

$$\sum a_i r_i = R. \tag{4.7.2}$$

The physical significance of the quantities r_i, R need not be prescribed, but we shall suppose that $r_i = r(q_i, p_i)$, so that the r_i are defined for each cell. By introducing a further Lagrange multiplier γ, the probability p for the gas condition $\{a_i\}$ can be maximized subject to the extra constraint and the results of section 3.5 amended accordingly. However, this amendment will be left as an exercise for the reader and, instead, we shall derive the new results by the more satisfactory method of mean values (section 4.1).

Thus, we first define a partition function

$$Z = z^{\varepsilon_1} w^{r_1} + z^{\varepsilon_2} w^{r_2} + \ldots, \tag{4.7.3}$$

where z, w are complex variables and sufficiently small units are chosen so that the ε_i and r_i can be taken to be integers. We then consider the multinomial expansion

$$Z^N = \sum \frac{N!}{a_1! \, a_2! \ldots} z^{\Sigma a_i \varepsilon_i} w^{\Sigma a_i r_i}, \tag{4.7.4}$$

the summation being over all sets $\{a_i\}$ satisfying the constraint $\Sigma a_i = N$. It follows that the coefficient of $z^U w^R$ in this expansion is the sum

$$\sum \frac{N!}{a_1! a_2! \ldots} \tag{4.7.5}$$

taken over all sets $\{a_i\}$ satisfying all three constraints. This must equal G, the number of members of the microcanonical ensemble modelling the system's state.

But the coefficient of z^U in the Taylor expansion about $z = 0$ of Z^N is

$$\frac{1}{2\pi i} \oint z^{-U-1} Z^N \, dz \tag{4.7.6}$$

and, for large U and N, we can approximate this by

$$\Phi = \{2\pi g_z(z,w)\}^{-1/2} z^{-U-1} Z^N, \tag{4.7.7}$$

where $z, g(z, w)$ satisfy

$$g \equiv \frac{U}{z} - N \frac{\partial}{\partial z} (\ln Z) = 0. \tag{4.7.8}$$

This determines z as a function of w which, as yet, is itself not fixed.

Φ is a function of w and can next be expanded in a Taylor series about $w = 0$; the coefficient of w^R in this expansion is

$$\frac{1}{2\pi i} \oint w^{-R-1} \Phi \, dw = \frac{z^{-U-1}}{2\pi i} \oint (2\pi g_z)^{-1/2} w^{-R-1} Z^N \, dw = G. \tag{4.7.9}$$

For large R and N, this is approximated by

$$(2\pi f_w)^{-1/2} z^{-U-1} (2\pi g_z)^{-1/2} w^{-R-1} Z^N = G, \tag{4.7.10}$$

where $w, f(w)$ are determined by the equations

$$f \equiv \frac{R}{w} - N \frac{\partial}{\partial w} (\ln Z) = 0. \tag{4.7.11}$$

The probability of the condition $\{a_i\}$ is given by

$$p = \frac{1}{G} \cdot \frac{N!}{a_1! a_2! \ldots} \tag{4.7.12}$$

and the mean value of a_1 over the possible gas conditions is accordingly

$$\bar{a}_1 = \sum p a_1 = \frac{N}{G} \sum \frac{(N-1)!}{a_0! a_2! \ldots}, \tag{4.7.13}$$

where $a_0 = a_1 - 1$. This last sum has to be calculated over all sets $\{a_0, a_2, \ldots\}$ satisfying the constraints

$$a_0 + a_2 + \ldots = N - 1, \quad a_0 \varepsilon_1 + a_2 \varepsilon_2 + \ldots = U - \varepsilon_1, \\ a_0 r_1 + a_2 r_2 + \ldots = R - r_1. \left.\right\} \tag{4.7.14}$$

Using the same method as that used to evaluate the sum (4.7.5), we calculate that

$$\bar{a}_1 = \frac{N}{G} \, (2\pi f_w)^{-1/2} \, z^{-U+\varepsilon_1-1} \, (2\pi g_z)^{-1/2} \, w^{-R+r_1-1} \, Z^{N-1}, \qquad (4.7.15)$$

where f, g, z and w are still given by equations (4.7.8) and (4.7.11), since ε_1, r_1, can be ignored by comparison with U and R respectively. Equations (4.7.10), (4.7.15) now yield the result

$$\bar{a}_1 = \frac{N}{Z} \, z^{\varepsilon_1} w^{r_1} = \frac{N}{Z} \exp\left(-\beta\varepsilon_1 - \gamma r_1\right), \qquad (4.7.16)$$

where we have put $z = e^{-\beta}, w = e^{-\gamma}$. This is the amended form of equation (4.1.13).

In general, if the a_i are subject to additional constraints $\Sigma a_i r_i = R$, etc., then

$$\bar{a}_i = \frac{N}{Z} \exp\left(-\beta\varepsilon_i - \gamma r_i - \ldots\right), \qquad (4.7.17)$$

where $\beta, \gamma, \ldots$ are determined by equations

$$\frac{\partial}{\partial\beta} \, (N \ln Z) = -U, \quad \frac{\partial}{\partial\gamma} \, (N \ln Z) = -R, \text{ etc.,} \qquad (4.7.18)$$

i.e. equations (4.7.8), (4.7.11), etc. The partition function Z will be formed from powers of complex variables $z, w, \ldots$ according to the pattern of equation (4.7.3).

The extension, as in section 4.2, to a mixture of gases is now straightforward, it being necessary to replace $(N \ln Z)$ by $(M \ln Z_1 + N \ln Z_2)$ in equations (4.7.18), where

$$Z_1 = \sum z^{\varepsilon_i} w^{r_i} \ldots, \quad Z_2 = \sum z^{\eta_j} w^{s_j} \ldots, \qquad (4.7.19)$$

the new constraints being of the form $\Sigma \, a_i r_i + \Sigma b_j s_j = R$, etc. The amended forms of equations (4.2.15) are

$$\bar{a}_i = \frac{M}{Z_1} \exp\left(-\beta\varepsilon_i - \gamma r_i - \ldots\right), \quad b_j = \frac{N}{Z_2} \exp\left(-\beta\eta_j - \gamma s_j - \ldots\right). \quad (4.7.20)$$

The analysis of a system placed in a large heat bath, when the 'system + bath' is subject to the additional constraints, now proceeds as in section 4.3. The amended form of equation (4.3.15) for the probability $\bar{n}_k$ of finding the system in the kth cell of its phase space is

$$\bar{n}_k = Z^{-1} \exp\left(-\beta e_k - \gamma r_k - \ldots\right), \qquad (4.7.21)$$

where

$$Z = \sum \exp\left(-\beta e_k - \gamma r_k - \ldots\right), \qquad (4.7.22)$$

and $\beta, \gamma, \ldots$ are determined by equations like (4.3.9) relating to the bath alone. In practice, we take β, γ, etc. to be given quantities, interpreting β in terms of the absolute temperature of the bath as usual by $\beta = 1/kT$.

Alternatively, the mean energy of the system in the bath is given by

$$U = \sum \bar{n}_k e_k = Z^{-1} \sum e_k \exp\left(-\beta e_k - \gamma r_k - \ldots\right)$$

$$= -\frac{1}{Z}\frac{\partial Z}{\partial \beta} = -\frac{\partial}{\partial \beta}(\ln Z) \tag{4.7.23}$$

and, by prescribing U, we can fix β through this equation.

Similarly, the mean $\bar{r}$ of the physical quantity r for the system is given by

$$\bar{r} = \sum \bar{n}_k r_k = -\frac{\partial}{\partial \gamma}(\ln Z). \tag{4.7.24}$$

Thus, γ can be fixed by prescribing $\bar{r}$.

4.8 Equipartition of energy. Law of Dulong and Petit

Consider an ideal gas for which a typical molecule has generalized coordinates q_r ($r = 1, 2, \ldots n$) and associated generalized components of momentum p_r. Let the molecule's energy be given by

$$\varepsilon = \alpha_1 p_1^2 + \alpha_2 p_2^2 + \ldots + \alpha_n p_n^2, \tag{4.8.1}$$

where the α_r are functions of the coordinates alone—in the absence of an external field of force, the expression for ε commonly takes this form (see. e.g. equation (4.5.4)). The molecular partition function now follows, thus:

$$Z = h^{-n} \sum \exp\left(-\beta \varepsilon_i\right) h^n = h^{-n} \int \exp\left(-\beta \varepsilon\right) dq_1 \ldots dq_n\, dp_1 \ldots dp_n. \tag{4.8.2}$$

Suppose we perform the integration with respect to p_1 by parts, to give

$$\int e^{-\beta \varepsilon} dp_1 = |p_1 e^{-\beta \varepsilon}| + \beta \int p_1 \frac{\partial \varepsilon}{\partial p_1} e^{-\beta \varepsilon} dp_1$$

$$= 2\beta \int \alpha_1 p_1^2 e^{-\beta \varepsilon} dp_1, \tag{4.8.3}$$

assuming that the integrated-out part vanishes at the limits of p_1 (often $-\infty$ and $+\infty$). It now follows that

$$Z = 2\beta h^{-n} \int \alpha_1 p_1^2 e^{-\beta \varepsilon} dq_1 \ldots dq_n\, dp_1 \ldots dp_n. \tag{4.8.4}$$

Similar identities can be proved by integration by parts with respect to p_2, $p_3, \ldots p_n$, successively. Addition of all these identities then leads to the result

$$nZ = 2\beta h^{-n} \int \varepsilon e^{-\beta \varepsilon} dq_1 \ldots dq_n\, dp_1 \ldots dp_n. \tag{4.8.5}$$

But, according to equation (4.3.16), the mean molecular energy is given by

$$U = -Z^{-1}\frac{\partial Z}{\partial \beta} = Z^{-1}h^{-n}\int \varepsilon e^{-\beta\varepsilon}\,dq_1 \ldots dp_n. \qquad (4.8.6)$$

Hence, using equation (4.8.5), we obtain

$$U = n/2\beta = \tfrac{1}{2}nkT. \qquad (4.8.7)$$

This result shows that each component of the molecule's energy contributes $\tfrac{1}{2}kT$ to the mean energy. This is the *Principle of Equipartition of Energy*. As an example, the diatomic gas studied in section 4.5 had molecules with five degrees of freedom corresponding to the coordinates (x, y, z, θ, ϕ); the mean energy per molecule should therefore be $5kT/2$, which is in agreement with equation (4.5.9).

The mean energy of a gas comprising N molecules is now found to be

$$U = \tfrac{1}{2}NnkT = \tfrac{1}{2}nvRT, \qquad (4.8.8)$$

where v is the mass of the gas in moles and R is the gas constant (see section 1.5). C_V, the heat capacity at constant volume for a gas, now follows thus:

$$C_V = \partial U/\partial T = \tfrac{1}{2}nvR. \qquad (4.8.9)$$

The value for C_V is usually quoted per mole, i.e. $\tfrac{1}{2}nR$ J K^{-1} mol^{-1}. Thus, for oxygen (O_2) at room temperature, $C_V = 21.0$ which agrees well with the classical theoretical value $5R/2 = 20.8$. However, the observed values at lower and higher temperatures deviate widely from this calculated value; at very low temperatures $C_V = 12.5 = 3R/2$, indicating that only the three degrees of freedom of the translatory motion are relevant—the suppression of the rotational motion can be explained by the quantum theory (section 7.3), but remained an anomaly for the classical theory; at high temperatures, C_V exceeds $5R/2$, the increment being due to an additional energy component arising from vibrations of the oxygen atoms along the line joining them (that this component makes negligible contribution at room temperature was also a mystery for the classical theory).

The principle of equipartition of energy can be extended to include the molecule's P.E. in an external field of force or due to internal forces, provided these take the quadratic form

$$\lambda_1 q_1^2 + \lambda_2 q_2^2 + \ldots + \lambda_n q_n^2. \qquad (4.8.10)$$

For then, integrating by parts with respect to q_1, we can prove as earlier that

$$Z = 2\beta h^{-n}\int \lambda_1 q_1^2\, e^{-\beta\varepsilon}\,dq_1 \ldots dp_n, \qquad (4.8.11)$$

together with similar identities for $q_2, \ldots, q_n$. Addition of these n identities, together with n identities like (4.8.4), then yields the result

$$2nZ = 2\beta h^{-n}\int \varepsilon e^{-\beta\varepsilon}\,dq_1 \ldots dp_n \qquad (4.8.12)$$

and it follows that

$$U = n/\beta = nkT. \tag{4.8.13}$$

Again, each of the $2n$ components of the molecular energy ε is found to contribute a mean energy $\frac{1}{2}kT$.

Boltzmann applied this principle to calculate the specific heat of a solid. Thus, consider a crystal whose atoms are all the same and are similarly situated. Each atom will interact with its neighbours, but it is reasonable to average out the combined influence of all the atoms of the crystal upon an individual atom and to represent it by a static field within which the atom vibrates about a position of equilibrium. Taking the equilibrium position as origin O of rectangular Cartesian coordinates (x, y, z), let $W(x, y, z)$ be the potential energy of the atom in this field. There is no loss of generality in assuming $W(0, 0, 0) = 0$. Further, W must have a minimum stationary point at O and, hence, $\partial W/\partial x = \partial W/\partial y = \partial W/\partial z = 0$ at this point. Hence, expanding W in a Taylor series about O, for small displacements from O, W will be given by the quadratic form

$$W = ax^2 + by^2 + cz^2 + 2fyz + 2gzx + 2hxy. \tag{4.8.14}$$

This quadratic form must be positive definite to give a minimum at O.

The Hamiltonian for the atom can now be written down as

$$H = \frac{1}{2m}(p_x^2 + p_y^2 + p_z^2) + W. \tag{4.8.15}$$

Following the usual method of analytical mechanics, we now rotate the axes about O (i.e. an orthogonal transformation of coordinates) to bring W to diagonal form and so re-express the Hamiltonian by

$$H = \frac{1}{2m}(p_x^2 + p_y^2 + p_z^2) + \frac{1}{2}m(\omega_x^2 x^2 + \omega_y^2 y^2 + \omega_z^2 z^2). \tag{4.8.16}$$

$(x, y, z,$ now refer to the new frame.) (x, y, z) are said to be *normal coordinates* for the atom and vibrations along the lines of the new axes are termed *normal vibrations*, the normal frequencies being $\omega_x/2\pi$, $\omega_y/2\pi$, $\omega_z/2\pi$.

We now treat the atoms of the crystal as independent systems, the energy of each being given by equation (4.8.16). Then, by the principle of equipartition, the mean energy for each atom will be $6 \cdot \frac{1}{2}kT = 3kT$. The heat capacity for the whole crystal is therefore $3Nk$ per degree K. If the crystal's mass in grammes is equal to the atomic weight m, then $N = L$ (Avogadro's number) and the heat capacity is $3kL = 3R$ (R = gas constant); this quantity is called the *atomic heat*. Taking $R = 8.3145 \, \mathrm{J\,K^{-1}}$, the theory indicates that the atomic heat should be about $25 \, \mathrm{J\,K^{-1}} = 6 \, \mathrm{cal\,K^{-1}}$. This is *Dulong and Petit's law*, obtained experimentally in 1819. Observed values (in calories) of atomic heats for various elements at $25\,°C$ are: aluminium 5.45, bismuth 4.66, calcium 5.82, copper 5.50, iron 5.63, silicon 4.42, silver 5.68—agreement with the predicted result is therefore quite good, but the discrepancies clearly call for further explanation (see section 6.2).

Treating a crystal of ordinary salt (NaCl) as a mixture of equal numbers of sodium and chlorine atoms arranged in a lattice structure, if N is the number of molecules, the sodium and chlorine atoms will each contribute a heat capacity $3Nk$, i.e. $6Nk$ in total. Thus, a mole of salt should have heat capacity $6kL = 6R$; this is its *molecular heat* 12 cal K^{-1} (the experimental value is 11.93). Crystals with triatomic molecules are similarly predicted to have a molecular heat of 18 cal K^{-1}.

4.9 Entropy and information

Consider, again, the canonical ensemble modelling a system possessing energy states e_i, immersed in a heat bath at temperature T. If p_i denotes the probability a member chosen at random from the ensemble is in the state e_i, then $p_i = Z^{-1} \exp(-\beta e_i)$ (equation (4.3.15)). It follows that

$$-k \sum_i p_i \ln p_i = kZ^{-1} \sum e^{-\beta e_i}(\beta e_i + \ln Z),$$
$$= kZ^{-1}(\beta \sum e_i e^{-\beta e_i} + Z \ln Z),$$
$$= k\beta U + k \ln Z,$$
$$= S, \tag{4.9.1}$$

having referred to equations (4.3.16) and (4.3.23).

This formula for the entropy, viz.

$$S = -k \sum p_i \ln p_i \tag{4.9.2}$$

provides another means of generalizing the concept of entropy to systems which are not in thermodynamical equilibrium, viz. we construct an ensemble to model the system's state and take p_i to be the proportion of the ensemble's members in the ith state. Some presentations of statistical mechanics (e.g. P. T. Landsberg, *Thermodynamics and Statistical Mechanics*, O.U.P., 1978) take this formula as the foundation on which the theory is constructed; in thermodynamical equilibrium, S is then assumed to be a maximum for admissible p_i (see Ex. 4, no. 6).

The formula also provides a link between statistical mechanics and information theory. In the latter theory, $-\ln p$ is taken as a measure of the information obtained when an event having probability p occurs. Thus, if the event is highly improbable, p is small and $-\ln p$ is very large, in agreement with our expectation that the event's happening greatly augments our stock of information. If, however, the event is known beforehand to be almost certain to occur (e.g. the sun's rising tomorrow), then $p = 1$ and $\ln p = 0$, showing that the actual occurrence of the event adds nothing to the information we already possess. In the light of these remarks, $-k \ln p_i$ is a measure of the information we acquire when the system is observed in the ith microstate; hence, by equation (4.9.2), S is a measure of the expected (or mean) information gained when we observe the microstate of a system selected at random from the ensemble. If the system is a highly organized one, we will already have detailed information regarding its microstate prior to any examination, and observation of the system will not greatly increase our knowledge in this respect; in this case, the entropy S will be low. If, however, the system is in a chaotic state, prior information as to its microstate will be very vague and an observation of the microstate must yield considerable new knowledge; S will now be large.

Consider the microcanonical ensemble as a special case. All members of this ensemble have the same energy and all are equally likely. Thus, if W is the number of members, $p_i = 1/W$ and $S = k \ln W$. We have, therefore, derived the Boltzmann formula (3.8.7) again and equation (4.9.2) can evidently be regarded as an extension of Boltzmann's principle.

Exercises 4

1. Obtain the equations (4.7.17) and (4.7.18) by maximizing the probability (4.7.12) subject to the constraints (3.5.3), (3.5.4) and (4.7.2).

2. Each molecule of a gas can be treated as a rigid body of revolution with principal moments of inertia (A, C, C) about its mass centre. If (θ, ϕ, ψ) are the Euler angles determining the orientation of the principal axes, obtain the Hamiltonian for the rotational motion in the form

$$ H = \frac{1}{2A} p_\psi^2 + \frac{1}{2C} \{ p_\theta^2 + (p_\phi - p_\psi \cos \theta)^2 \operatorname{cosec}^2 \theta \}. $$

Hence show that the partition function for the molecule moving in a container of volume V is given by

$$ Z = 64 h^{-6} \pi^5 V A^{1/2} C m^{3/2} (kT)^3 $$

and deduce that the mean energy per molecule is doubled by the rotational motion.

3. For a crystal modelled as a collection of N weakly interacting atoms, each capable of existing in two energy states $E, E + \varepsilon$ (see sections 3.8 and 4.4), show that its entropy at temperature T is given by

$$ S = Nk\{ \ln(1 + e^{-x}) + x/(1 + e^x) \}, $$

where $x = \beta\varepsilon = \varepsilon/kT$. Verify that $S \to 0$ as $T \to 0$.

4. A certain system comprises N weakly interacting subsystems, each of which has four possible energy states, one of energy E, two of energy $E + \varepsilon$ and one of energy $E + 2\varepsilon$. Calculate the partition function for a subsystem and use the first of equations (4.4.7) to deduce the mean energy of the system at temperature T. Hence show that the heat capacity C_V ($= \partial U/\partial T$) of the system is

$$ 2Nk \frac{x^2 e^x}{(e^x + 1)^2}, $$

where $x = \varepsilon/kT$. Verify that C_v is a maximum at $T = 3\varepsilon \times 10^{22}$ K approximately (ε in joules).

5. Show that the partition function for the model of a paramagnetic material analysed in section 4.6 is Z^N, where

$$ Z = 8\pi^2 C \mu_0 m H h^{-2} \alpha^{-2} \sinh \alpha $$

where $\alpha = \mu_0\ mH\beta$. Deduce that the energy and entropy of the material are given by

$$U = N\mu_0\ mH\left(\frac{2}{\alpha} - \coth\alpha\right),$$

$$S/(Nk) = 2 + \ln(8\pi^2 C\mu_0\ mHh^{-2}) + \ln(\alpha^{-2}\sinh\alpha) - \alpha\coth\alpha,$$

respectively. Show that $S \to -\infty$ as $T \to 0$ (thus, the model is not consistent with the third law (see section 5.5)).

6. The energy of a certain system is e_i with probability p_i. If the mean energy of the system is known to be U and taking its entropy S to be given by equation (4.9.2), show that S is maximized with respect to the variables p_i, subject to the further constraint $\Sigma\, p_i = 1$, when the distribution is canonical.

CHAPTER 5

Quantum statistics

5.1 Quantization of phase space

Suppose a gas molecule is prepared in a prescribed pure quantum state ψ by the observation of the values of a maximal set of mutually compatible observables. Immediately afterwards, its x-coordinate is observed to have the value x. If this procedure is carried out a very large number of times (i.e. preparation in state ψ and subsequent measurement of x), unless x happens to be one of the observables used to define ψ (i.e. unless ψ is an eigenstate of x), the observed values of x will not all be identical, but will be spread about some mean value $\bar{x}$. Let Δx be the standard deviation measuring the extent of this departure from the mean. Next, suppose the x-component of linear momentum p_x of the molecule is similarly observed a large number of times for the state ψ and let Δp_x be the standard deviation for the sequence of values obtained. Then, it can be proved to be a consequence of the basic principles of quantum theory that

$$\Delta x \, \Delta p_x \geqslant h/4\pi, \tag{5.1.1}$$

where h is *Planck's constant* (6.626×10^{-34} Js). This is *Heisenberg's Uncertainty Principle*.

If, therefore, we construct a phase space by setting up rectangular Cartesian axes Oxp_x in a plane and try to represent the state of the molecule by a point with coordinates (x, p_x), the position of this point will be uncertain, but can be expected, with high probability, to lie with its x-coordinate in the range $(\bar{x} - 2\Delta x, \bar{x} + 2\Delta x)$ and its p_x-coordinate in the range $(\bar{p}_x - 2\Delta p_x, \bar{p}_x + 2\Delta p_x)$, i.e. within a rectangle of area $16\Delta x \, \Delta p_x$. According to the inequality (5.1.1), this area cannot be smaller than $4h/\pi$, i.e. approximately h.

The y- and z-coordinates of the molecule and the corresponding components of momentum can be treated similarly and we conclude that the point representing the molecule's state in its six-dimensional phase space can be expected to lie in a cell of volume h^3.

More generally, if the gas comprises N molecules, the point representing the microscopic state of the gas can be taken to lie in a cell of the gas's phase space having volume h^{3N}.

If we transform from Cartesian coordinates and their associated components of momentum to generalized coordinates q_i and their corresponding momenta p_i, it may be proved (see e.g. R. C. Tolman, *Principles of Statistical Mechanics*, Chap. III) that this transformation preserves volumes of corresponding regions of phase

100

space invariant. Thus, the results we have just found apply in any phase space used to represent the molecular and gaseous states.

It is clear that the uncertainty principle implies that a system's phase space is quantized, such that the smallest volume it is possible to accept for a cell is of the order of h^n, where n is the number of degrees of freedom of the system (i.e. number of generalized coordinates). Our notation in Chapter 3 is in conformity with this result.

5.2 Density matrix

The first step towards constructing a statistical quantum theory is to identify the counterpart of the phase point density function ρ, the conservation of whose uniform distribution over phase space suggested the principle of randomicity in phase upon which the classical theory is founded. It will be shown in section 5.3 that the required object is the density matrix; in this section, we shall define this concept and establish its principal properties.

In the classical statistics, $\rho(q, p, t)$ specified the density over phase space of the points representing the states of the various members of the ensemble being used to model a statistical system at time t. Thus, we are looking for a quantity which will serve to describe a whole ensemble of quantum states. As in the classical theory, each such state will be supposed specified to the maximum extent permitted by the underlying fundamental dynamical principles—in this case, the principles of quantum mechanics. Each member of the ensemble will therefore be assumed established in a pure state described by a state vector $\psi(t)$, which will be a column matrix $(\psi_1, \psi_2, \ldots)^T$ (T denotes transpose) if a discrete quantum representation is being used, or a wave function $\psi(q, t)$ if a continuous representation based on a maximal compatible set of continuous observables $\{q\} = \{q_1, q_2, \ldots\}$ is adopted (mixed representations are also acceptable and it is assumed the reader is proficient in the art of interpreting formulae to suit all possibilities; however, in the subsequent argument, to fix ideas, it will be convenient to assume that a discrete representation is being employed, so that ψ, ψ^T are column and row matrices respectively). Suppose the ensemble comprises g_1 systems in a state ψ_1, g_2 systems in a state ψ_2, and so on, there being $G = \Sigma g_r$ systems in the ensemble altogether. Then the probability that the actual system is in the state ψ_r is $P_r = g_r/G$. The *density matrix* for the ensemble is then defined by the equation

$$\rho = \sum_r P_r \psi_r \psi_r^\dagger, \qquad (5.2.1)$$

where $\psi_r^\dagger$ is the complex conjugate transpose of ψ_r. If ψ_r is an $N \times 1$ matrix, then $\psi_r^\dagger$ is a $1 \times N$ matrix and ρ is clearly an $N \times N$ matrix. (We prefer this notation to the cumbersome and ugly $|\psi_r\rangle$ (for ψ_r) and $\langle\psi_r|$ (for $\psi_r^\dagger$).)

First, note that ρ is Hermitian, for

$$\rho^\dagger = \sum_r P_r^* (\psi_r \psi_r^\dagger)^\dagger = \sum_r P_r \psi_r \psi_r^\dagger = \rho, \qquad (5.2.2)$$

P_r being real. (Asterisks indicate complex conjugates.)

Next, if ψ_{nr} is the element in the nth row of ψ_r, then

$$\rho_{mn} = \sum_r P_r \psi_{mr} \psi_{nr}^* \qquad (5.2.3)$$

is the element in the mth row and nth column of ρ. In particular

$$\rho_{nn} = \sum_r P_r \psi_{nr} \psi_{nr}^* = \sum_r P_r |\psi_{nr}|^2 = \sum_r P_r p_{nr}, \qquad (5.2.4)$$

where $p_{nr} = |\psi_{nr}|^2$ is the probability a system in the state ψ_r will be observed in the nth eigenstate of the orthonormal basis for the discrete representation being used. It follows that the diagonal element ρ_{nn} of ρ is the overall probability of finding a system, chosen at random from the ensemble, in this nth eigenstate of the basis. Clearly ρ_{nn} must be a positive real number and

$$\sum_n \rho_{nn} = \sum_r P_r \sum_n p_{nr} = \sum_r P_r = 1, \qquad (5.2.5)$$

as we would expect. This sum of diagonal elements is termed the *trace* of ρ and we write $\Sigma \rho_{nn} = \mathrm{Tr}\rho = 1$.

Let x be any observable of the system. Then its mean or expected value in the quantum state ψ will be denoted by $\langle x \rangle$; quantum theory requires that

$$\langle x \rangle = (\psi, x\psi) = \psi^\dagger x\psi, \qquad (5.2.6)$$

where x also stands for the $N \times N$ matrix representing the observable in our discrete representation. (Note: $(\psi, \phi) = $ scalar product of ψ and $\phi = \psi^\dagger \phi$.) The mean value of x over the ensemble is now calculated to be

$$
\begin{aligned}
\bar{x} &= \sum_r P_r \langle x \rangle_r = \sum_r P_r \psi_r^\dagger x\psi_r = \sum_r P_r \sum_{m,n} \psi_{mr}^* x_{mn} \psi_{nr} \\
&= \sum_{m,n} x_{mn} \sum_r P_r \psi_{nr} \psi_{mr}^* = \sum_{m,n} x_{mn} \rho_{nm} \\
&= \mathrm{Tr}\,(x\rho) = \mathrm{Tr}\,(\rho x), \qquad (5.2.7)
\end{aligned}
$$

having used equation (5.2.3).

This result (5.2.7) shows that, once the density matrix for an ensemble has been calculated, the mean values of all system observables can be found. But, it is these mean values which we shall later identify (as in the classical theory) with the observed values of the system's macroscopic variables (parameters of state). It follows, therefore, that two ensembles having the same density matrix are *statistically equivalent*, i.e. lead to the same predicted values of all macroscopic observables, and it is not necessary for our purpose to distinguish between them— they model the same statistical situation. This justifies our use of the density matrix to provide a comprehensive description of an ensemble.

Now suppose we transform from one discrete representation to another. The effect on a state vector ψ is to subject it to a unitary transformation, viz.

$$\psi' = u\psi, \qquad (5.2.8)$$

where u is an $N \times N$ unitary matrix (i.e. $uu^{\dagger} = u^{\dagger}u = I_N$). In the 'dashed' representation, the density matrix for our ensemble is ρ', where

$$\rho' = \sum P_r \psi'_r \psi'^{\dagger}_r = \sum P_r u\psi_r \psi^{\dagger}_r u^{\dagger} = u \left(\sum P_r \psi_r \psi^{\dagger}_r \right) u^{\dagger}$$
$$= u\rho u^{\dagger}. \tag{5.2.9}$$

This is the transformation equation for density matrices.

It is known that any Hermitian matrix ρ can be brought to diagonal form by a unitary transformation of this type, the diagonal elements being the eigenvalues of ρ (all real and positive for a matrix of this type), which we shall denote by $\{\rho_1, \rho_2, \ldots\}$. Suppose this has been done, so that

$$\rho = \text{diag}(\rho_1, \rho_2, \ldots, \rho_N). \tag{5.2.10}$$

Defining e_r $(r = 1, 2, \ldots, N)$ to be the $N \times 1$ column matrix with all its elements zero, except the rth which is unity, i.e.

$$e_{nr} = \delta_{nr} \tag{5.2.11}$$

(where δ_{nr} is the Kronecker delta symbol), we can express the density matrix (5.2.10) thus:

$$\rho = \sum_r \rho_r e_r e^{\dagger}_r. \tag{5.2.12}$$

In an arbitrary representation, let $\alpha_1, \alpha_2, \ldots, \alpha_N$ denote the set of vectors forming the orthonormal basis of the representation for which ρ is in diagonal form. Then, in this latter representation, $\alpha_r = e_r$ and equation (5.2.12) can be written

$$\rho = \sum_r \rho_r \alpha_r \alpha^{\dagger}_r. \tag{5.2.13}$$

In this form, the equation is valid in any representation and it shows that the ensemble is statistically equivalent to one comprising $G\rho_1$ members in the state α_1, $G\rho_2$ members in the state α_2, etc., where the $\{\alpha_r\}$ form a complete orthonormal set. From equation (5.2.5), it follows that

$$\sum \rho_r = 1, \tag{5.2.14}$$

so that the total number of members of the ensemble is G, quite correctly.

To summarize, there is no loss of generality in assuming that the distinct states of the systems of an ensemble form a complete orthonormal set $\{\alpha_r\}$, in which case the probabilities of finding the system in the states $\alpha_1, \alpha_2, \ldots$ are the eigenvalues of ρ, viz. $\rho_1, \rho_2, \ldots$. The complete orthonormal set $\{\alpha_r\}$ will invariably be chosen by augmenting the energy observable for the system by a sufficient number of other compatible observables to form a maximal compatible set C and then defining the $\{\alpha_r\}$ to be the normalized eigenvectors of C. Note, therefore, that every member of the ensemble will have a sharp energy; the other observables which are to be sharp can be selected to suit our convenience in each case.

As we shall see, in the quantum theory, the orthonormal states α_r play an analogous role to that of the states determined by the cells in the classical theory.

5.3 Liouville's theorem

If H is the Hamiltonian matrix for the system being modelled by an ensemble of systems in pure states ψ_1, ψ_2, etc., then each of these states will vary with the time t as determined by the Schrödinger equation

$$i\hbar \frac{\mathrm{d}\psi_r}{\mathrm{d}t} = H\psi_r. \tag{5.3.1}$$

Taking the conjugate transpose of both members of this equation, we obtain

$$-i\hbar \frac{\mathrm{d}\psi_r}{\mathrm{d}t} = \psi_r^\dagger H^\dagger = \psi_r^\dagger H, \tag{5.3.2}$$

since H is always Hermitian. Then, differentiating equation (5.2.1), we find

$$\begin{aligned}
i\hbar \frac{\mathrm{d}\rho}{\mathrm{d}t} &= i\hbar \sum P_r \left(\frac{\mathrm{d}\psi_r}{\mathrm{d}t} \psi_r^\dagger + \psi_r \frac{\mathrm{d}\psi_r^\dagger}{\mathrm{d}t} \right) \\
&= \sum P_r (H\psi_r \psi_r^\dagger - \psi_r \psi_r^\dagger H) \\
&= H\rho - \rho H = [H, \rho],
\end{aligned} \tag{5.3.3}$$

where $[H, \rho]$ is called the *commutator* of H and ρ.

Equation (5.3.3) is the quantum mechanical analogue of the classical equation (3.3.5) and justifies our regarding the density matrix ρ as corresponding, in the new theory, to the phase point density ρ of the old theory.

Having established the quantum form of Liouville's theorem, it is possible to identify types of ensemble which remain in statistical equilibrium. These are ensembles for which the solution of equation (5.3.3) is a constant matrix ρ; then, $\mathrm{d}\rho/\mathrm{d}t = 0$ and ρ must commute with H if equation (5.3.3) is to be satisfied.

Consider an ensemble whose members are in the N orthonormal states α_1, α_2, etc. at some instant, the numbers in each state being all the same. Then, $P_r = 1/N$ and the density matrix is

$$\rho = \frac{1}{N} \sum \alpha_r \alpha_r^\dagger. \tag{5.3.4}$$

As always, we shall assume the set $\{\alpha_r\}$ to be complete and orthonormal, so that $(\alpha_r, \alpha_s) = \alpha_r^\dagger \alpha_s = \delta_{rs}$. Then

$$\rho \alpha_s = \frac{1}{N} \sum \alpha_r \alpha_r^\dagger \alpha_s = \frac{1}{N} \sum \alpha_r \delta_{rs} = \frac{1}{N} \alpha_s, \tag{5.3.5}$$

showing that $\{\alpha_s\}$ is a complete set of eigenvectors for ρ, the corresponding eigenvalues being all $1/N$. Hence, in a representation for which ρ is diagonal, we shall have

$$\rho = N^{-1} I_N, \tag{5.3.6}$$

where I_N is the unit matrix of order N. In any other representation, by equation

(5.2.9), ρ will transform to

$$\rho' = N^{-1} u I_N u^\dagger = N^{-1} u u^\dagger = N^{-1} I_N. \tag{5.3.7}$$

Thus, for such an ensemble, $\rho = N^{-1} I_N$ in every representation.

But $\rho = N^{-1} I_N$ satisfies equation (5.3.3) and it follows that the ensemble is in statistical equilibrium. It is the analogue of the classical uniform ensemble (see section 3.3) and is called by the same name. Any system modelled by this ensemble has the same probability $1/N$ of being found in every one of the base states α_r. The circumstance that this ensemble is in statistical equilibrium indicates that the principles of quantum mechanics alone do not give us any reason for attaching greater weight to one of the states $\{\alpha_r\}$ rather than another; thus, in the absence of information, the same probability should be allocated to each of the base states $\{\alpha_r\}$. This is the quantum counterpart of the principle of randomicity in phase (section 3.3).

Suppose that, for a certain system, $a, b, \ldots$ are compatible observables whose values are given at some instant t^0. Augment these observables by others which are compatible with them to form a maximal compatible set C. Let $\alpha_1, \alpha_2, \ldots, \alpha_N$ be a complete orthonormal set of eigenvectors for C, α_r $(r = 1, 2, \ldots, M)$ being the eigenstates in which the observables $a, b, \ldots$ take the given values. According to the principle just explained, probability $1/M$ should be allocated to each of the states $\alpha_r (r = 1, 2, \ldots, M)$ and zero probability to the remaining states of the complete set. Thus, the density matrix for the system at time t^0 should be taken to be

$$\rho = \frac{1}{M} \sum_{r=1}^{M} \alpha_r \alpha_r^\dagger. \tag{5.3.8}$$

In general, the constant matrix ρ given by the last equation will not commute with H and hence will not satisfy equation (5.3.3). In this case, the system will not remain in the statistical state specified by ρ. If, however, one of the observables $a, b, \ldots$ is the energy and this takes the sharp value e, then we must have

$$H\alpha_r = e\alpha_r, \quad \alpha_r^\dagger H = e\alpha_r^\dagger, \tag{5.3.9}$$

for $r = 1, 2, \ldots, M$. Then,

$$H\rho - \rho H = \frac{1}{M} \sum_{r=1}^{M} (H\alpha_r \alpha_r^\dagger - \alpha_r \alpha_r^\dagger H)$$

$$= \frac{1}{M} \sum_{r=1}^{M} (e\alpha_r \alpha_r^\dagger - \alpha_r e\alpha_r^\dagger) = 0 \tag{5.3.10}$$

and the commutator $[H, \rho]$ vanishes. In these circumstances, therefore, the system will be in statistical equilibrium with density matrix given by equation (5.3.8) for all t. The energy taking a known value e, this type of ensemble is the quantum analogue of the classical microcanonical ensemble.

The argument leading to equation (5.3.8) can be generalized. Suppose that the values of the compatible observables $a, b, \ldots$ are not known precisely at the time

t^0, but that probabilities p_1, p_2, etc. can be assigned to the sets of eigenvalues $(a_1, b_1, \ldots)$, $(a_2, b_2, \ldots)$, etc. respectively. We again augment these observables to construct a maximal set C and so generate a complete orthonormal set of eigenvectors of C. Of these eigenvectors, let $\alpha_r^{(1)}$ ($r = 1, 2, \ldots, M_1$) determine states in which a, b, etc. take values $(a_1, b_1, \ldots)$, let $\alpha_s^{(2)}$ ($s = 1, 2, \ldots, M_2$) determine states in which $a, b, \ldots$ take values $(a_2, b_2, \ldots)$, and so on. Then, the system will be found in each of the states $\alpha_r^{(1)}$ with probability p_1/M_1 (i.e. probability p_1 its state vector belongs to this group, and probability $1/M_1$ its state vector is a specific member of the group); similarly, the system will be found in each of the states $\alpha_s^{(2)}$ with probability p_2/M_2, and so on. We can now construct the density matrix at t^0 in the form

$$\rho = \frac{p_1}{M_1} \sum_{r=1}^{M_1} \alpha_r^{(1)} \alpha_r^{(1)\dagger} + \frac{p_2}{M_2} \sum_{s=1}^{M_2} \alpha_s^{(2)} \alpha_s^{(2)\dagger} + \ldots \qquad (5.3.11)$$

If one of the observables $a, b, \ldots$ is the energy, then

$$H\alpha_r^{(1)} = e_1 \alpha_r^{(1)}, \quad H\alpha_s^{(2)} = e_2 \alpha_s^{(2)}, \text{ etc.}, \qquad (5.3.12)$$

where e_i is the energy eigenvalue taken from the set $(a_i, b_i, \ldots)$, and it follows as before that ρ commutes with H and the ensemble will be in statistical equilibrium. Such a density matrix is the quantum analogue of the classical phase point density which is a function of constants α, β, $\ldots$ for the system's motion (equation (3.3.9)). A special case is when the set $a, b, \ldots$ reduces to the energy observable alone and the energy is known to be sharp but only the probability distribution of its eigenvalue is known; if $p_i \propto \exp(-\beta e_i)$, the corresponding ensemble is canonical (cf. the classical canonical ensemble, section 4.3).

5.4 Quantum interpretation of classical theory of molecular aggregates

Consider a system comprising N molecules which are free to vibrate in an external field of force *at known locations*. Assuming that the energy of interaction between the molecules is negligible, if H_n is the Hamiltonian for the motion of the nth molecule in the external field, then the Hamiltonian for the complete system is H, where

$$H = H_1 + H_2 + \ldots + H_N, \qquad (5.4.1)$$

i.e. no interaction energy terms need be included. If, now, the nth molecule is in a state with energy eigenvalue ε_n, its state vector ψ_n will satisfy

$$H_n \psi_n = \varepsilon_n \psi_n, \qquad (5.4.2)$$

provided the influence of adjacent molecules can be ignored (as we are supposing). The state vector for the whole system is then the Kronecker (or outer) product of the molecular states, viz. $\psi = \psi_1 \psi_2 \ldots \psi_N$, and we find that

$$H\psi = (H_1 \psi_1)\psi_2 \ldots, \psi_N + \psi_1 (H_2 \psi_2) \ldots \psi_N + \ldots + \psi_1 \psi_2 \ldots (H_N \psi_N)$$
$$= (\varepsilon_1 + \varepsilon_2 + \ldots + \varepsilon_N) \psi_1 \psi_2 \ldots \psi_N = U\psi, \qquad (5.4.3)$$

where
$$U = \varepsilon_1 + \varepsilon_2 + \ldots + \varepsilon_N. \tag{5.4.4}$$

(Note: H_n only operates on the state vector ψ_n.) Thus, the system's energy is the sum of the molecular energies, exactly as was assumed in section 3.5.

The argument of section 3.5 can now be adapted to the present circumstances. Instead of taking ε_i to be the molecular energy associated with the ith cell of the molecular phase space, we define it to be the energy eigenvalue for the ith member of a complete orthonormal set of molecular states (all eigenstates of the energy). A complete orthonormal set of *system* states is now generated by permitting the molecules to range independently over their orthonormal states and, according to the principle enunciated in the last section, in the absence of further information, these system states all have the same probability and each is included the same number of times in the Gibbs ensemble; for simplicity, we take one of each. Now suppose the total energy U is known. Then we identify those members of the ensemble for which a_1 molecules are in the energy eigenstate ε_1, a_2 are in the eigenstate ε_2, etc., where the integers $\{a_i\}$ are required to satisfy the constraints

$$\sum a_i = N, \quad \sum a_i \varepsilon_i = U. \tag{5.4.5}$$

Each such set $\{a_i\}$ is said to determine a *condition* of the system. The number of members of the ensemble modelling the condition is given by the expression (3.5.1) and the probability for the condition is accordingly given by equation (3.5.2). The argument now proceeds as before and the Maxwell–Boltzmann law is found to give the numbers of molecules in the various energy eigenstates.

It should here be noted that our argument is invalidated if the molecules are not localized, but are free to move over the interior of a container as a gas. This is because, in these circumstances, the molecules are indistinguishable and, according to quantum principles, it is incorrect to regard the condition $\{a_i\}$ as being separable into the number of distinct states given by the expression (3.5.1). If molecules A and B are in the states ε_A, ε_B respectively, the state of the complete system is unaltered if we take A to be in the state ε_B and B to be in state ε_A. Thus, all the states we have earlier regarded as distinguishable and therefore as contributing separately to the condition $\{a_i\}$ represent only one state of the gas and the probability for this condition is $1/K$, where K is the number of *conditions*, and not as given by equation (3.5.2).

Provided the molecules are localized, therefore, our system can be analysed by the method of mean values (section 4.1) and then the extension to a mixture of localized molecules proceeds as in section 4.2.

Now suppose we have a quantum system whose energy eigenvalues are e_1, e_2, etc. In general, these energy eigenvalues will be degenerate, but by adjoining to the energy observable a sufficient number of other compatible observables, we shall be able to define a complete orthonormal set of eigenstates, whose energy eigenvalues we shall again denote by e_k ($k = 1, 2, \ldots$), although the e_ks will not now be all different. In the absence of further information, all these eigenstates must be regarded as being equally probable. If this system is placed in a heat bath comprising a mixture of *localized molecules*, the argument of section 4.3 can be applied and the conclusions reached in that section regarding the state of the system at temperature T (equations (4.3.15) and (4.3.16)) are therefore valid. Since these results are not affected by the characteristics of the heat bath, we shall assume them to be valid for any type of heat bath or environment at temperature T. (Note: Although the heat bath was specialized to comprise localized molecules, the system placed in the bath is not limited in any way, and may (as in chapter 7) be taken to be a gas of indistinguishable molecules.)

Thus, to calculate the properties of *any* quantum system in statistical equilibrium at temperature T, it is only necessary to compute its partition function (4.3.5) and then to apply the results of section 4.3. If the system can be separated into a number of non-interacting subsystems with partition functions Z_1, Z_2, etc. then, *provided the systems are distinguishable*, the argument given in section 4.4 is valid and the partition function for the complete system is given by $Z = Z_1 Z_2 \ldots$. If, however, two of the subsystems are indistinguishable, the energy state in which the subsystems have energies ε, η respectively is not to be counted as different from that in which the subsystems have energies η, ε when calculating the partition function and the argument of section 4.4 accordingly breaks down. The analysis of such systems will be taken up in Chapter 7.

5.5 Absolute zero. Third law of thermodynamics

Suppose we arrange the energy eigenvalues e_i of a quantum system in increasing order of magnitude, taking the first m to be equal ($e_1 = e_2 = \ldots = e_m = \varepsilon$), the next n to be equal ($e_{m+1} = e_{m+2} = \ldots = e_{m+n} = E$), and so on. Then $\varepsilon < E$ and the partition function is given by

$$Z = me^{-\beta\varepsilon} + ne^{-\beta E} + \ldots, \tag{5.5.1}$$

the terms indicated by dots tending to zero more rapidly than $e^{-\beta E}$ as $\beta \to +\infty$ (i.e. $T \to 0$).

The internal energy of the system is now given by equation (4.3.16) as

$$U = \frac{1}{Z}(m\varepsilon e^{-\beta\varepsilon} + nEe^{-\beta E} + \ldots)$$

$$= \varepsilon\left[1 + \frac{nE}{m\varepsilon}e^{-\beta(E-\varepsilon)} + \ldots\right]\left[1 + \frac{n}{m}e^{-\beta(E-\varepsilon)} + \ldots\right]^{-1}$$

$$= \varepsilon + \frac{n}{m}(E - \varepsilon)e^{-\beta(E-\varepsilon)} + \ldots, \tag{5.5.2}$$

the expansions being valid for sufficiently large β.

Referring to equations (4.3.23), we can now derive a similar expansion for the entropy S, viz.

$$S = k\ln\left[me^{-\beta\varepsilon}\left\{1 + \frac{n}{m}e^{-\beta(E-\varepsilon)} + \ldots\right\}\right] + k\beta\left[\varepsilon + \frac{n}{m}(E-\varepsilon)e^{-\beta(E-\varepsilon)} + \ldots\right]$$

$$= k\ln m + \frac{kn}{m}\{1 + \beta(E-\varepsilon)\}e^{-\beta(E-\varepsilon)} + \ldots \tag{5.5.3}$$

Letting $\beta \to +\infty$ so that $T \to 0$, we see that

$$S \to k \ln m. \tag{5.5.4}$$

The ground states of quantum systems are almost invariably non-degenerate, so that $m = 1$ and, then, $S \to 0$. In any case, m can be expected always to be small and the limit for S, as absolute zero is approached, can therefore be accepted as being effectively zero in all circumstances.

As has already been pointed out, S was originally so defined as to be arbitrary to the extent of an additive constant (see equation (1.7.4)). Our derivation of equation (4.3.23) established that this constant is independent of T and the e_ks. This constant was ignored in equation (4.3.23), but had it been retained, the present result would have shown it to be the universal value of S at $T = 0$ for all systems. Thus, ignoring the constant in equation (4.3.23) amounts to putting this universal constant to zero and accepting the state of any system at $T = 0$ as the datum from which S is measured.

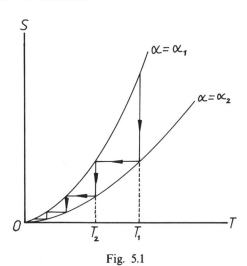

Fig. 5.1

The convergence of the entropy for all systems in all states to the same constant value as absolute zero is approached, implies that it is impossible actually to reach $T = 0$ by a finite sequence of manipulations of any thermodynamic system (i.e. the third law). Thus, the recognized method of achieving low temperatures is first to cool the system to a temperature T_1 by immersing it in a reservoir at this temperature and then to vary a parameter of state α from α_1 to α_2 isothermally. The second stage is to isolate the system and then to vary the parameter's value from α_2 back to α_1 adiabatically, thus lowering the temperature of the system to T_2 (e.g. adiabatic demagnetization, see section 2.5). If this process is performed on a large number of such systems, these can then be employed as a reservoir and the procedure repeated to lower the temperature of a further specimen system still further. However, a consideration of the system's characteristics $S = S(T, \alpha)$, α = constant, in the ST-plane, makes it clear that a finite number of operations of this type will never bring the system to a state in which $T = 0$, since all these curves have to meet at the origin $S = T = 0$ (see Fig. 5.1).

5.6 Grand canonical ensemble

Thus far, we have been concerned with systems which exchange heat with their surroundings, but whose constituent particles are conserved. In this section, the system under analysis will be permitted to lose and gain particles from its environment in the manner in which, for example, vapour evaporates and condenses from water exposed to the atmosphere. However, by defining our system to be the portion of a gas located in a specified spatial region V, even a gas in a closed container can be treated on this basis; the portion of the gas occupying the region V is surrounded by the rest of the gas in the container and this behaves as its environment with which it exchanges heat and molecules. This approach will be found very convenient when we develop the theory of the quantal gas in Chapter 7. In any case, it will always be assumed that the system has achieved its equilibrium state, so that the rates at which it gains and loses particles averaged over a long time period are equal.

Let M denote the number of particles (assumed identical) which are free to move between our system S and its environment. Thus, at any time, S comprises some of these particles together, possibly, with other particles which are regarded as anchored in S and are never exchanged with the environment; the former particles will be classed as being of type A and the latter of type B. When S contains n particles of type A, let e_{kn} $(k = 1, 2, \ldots; n = 0, 1, \ldots, M)$ denote its energy eigenvalues (repeated as often as necessary to allow for degeneracy). As our model for the environment, we shall take this to comprise a very large number of localized particles of type A, whose interactions will be assumed to be negligible; here we are appealing to physical intuition that the form of the conclusions we shall reach in regard to the probable state of S are not affected by the nature of the environment. The system S, together with its environment, will be regarded as isolated and to have total energy E. The total number of particles of type A in this closed system will be denoted by R $(\gg M)$. Thus, the actual number q of type A particles in the environment varies between $(R - M)$ and R, as the number in S varies between M and 0.

The energy eigenstates for each localized particle of the environment will be denoted by ε_i $(i = 1, 2, \ldots)$ and a condition of the environment will be specified by stating the number a_i of particles in each state ε_i. Then

$$\sum a_i = q \tag{5.6.1}$$

and the number of equally probable states capable of modelling this condition is

$$\frac{q!}{a_1! \, a_2! \cdots}. \tag{5.6.2}$$

To specify the state of S, we introduce the numbers n_k $(k = 1, 2, \ldots)$, where $n_k = 0$ if S is not in the state e_{kn} and $n_k = 1$ if it is.

The energy constraint can now be written

$$\sum_i a_i \varepsilon_i + \sum_k n_k e_{kn} = E, \qquad (5.6.3)$$

where the energy unit is chosen so that ε_i, e_{kn} and E are all integers (E very large). There is also the constraint

$$q + n = R, \qquad (5.6.4)$$

where n can vary from 0 to M.

An admissible condition of the whole system (S + environment), can now be specified by giving a set of integral values for $\{a_i, n_k, n\}$ satisfying the constraints (5.6.1), (5.6.3). The number of members of the Gibbs ensemble modelling this condition is given at (5.6.2) and the total number of members of the ensemble is therefore

$$G = \sum \frac{q!}{a_1! \, a_2! \cdots} \qquad (5.6.5)$$

summed over all admissible sets $\{a_i, n_k, n\}$.

We next define a partition function for a particle of the environment, viz.

$$Z = \sum_i z^{\varepsilon_i}. \qquad (5.6.6)$$

The partition function Z_n for S depends upon the number n of type A particles it contains; we shall write

$$Z_n = \sum_k z^{e_{kn}}. \qquad (5.6.7)$$

Expanding $Z^q Z_n$ in powers of z, we get

$$Z^q Z_n = \sum \frac{q!}{a_1! \, a_2! \cdots} z^{\sum a_i \varepsilon_i + \sum n_k e_{kn}}, \qquad (5.6.8)$$

where the summation is to be carried out over all sets $\{a_i, n_k\}$ satisfying the constraints

$$\sum_i a_i = q, \quad \sum_k n_k = 1. \qquad (5.6.9)$$

Summing the identity (5.6.8) with respect to n from 0 to M, we get

$$X = \sum_n Z^q Z_n = \sum \frac{q!}{a_1! \, a_2! \cdots} z^{\sum a_i \varepsilon_i + \sum n_k e_{kn}}, \qquad (5.6.10)$$

where the summation in the right-hand member is now carried out over all sets $\{a_i, n_k, n\}$ satisfying the constraints (5.6.9) and $0 \leqslant n \leqslant M$.

It now follows that the coefficient of z^E in the expansion of X is the number G given by equation (5.6.5). This can be calculated from a contour integral thus:

$$G = \frac{1}{2\pi i} \oint z^{-E-1} X \, dz = \frac{1}{2\pi i} \sum_n \oint z^{-E-1} Z^q Z_n \, dz. \qquad (5.6.11)$$

Making use, once again, of the result found in Appendix C, we obtain

$$\frac{1}{2\pi\iota}\oint z^{-E-1}\,Z^q Z_n\,\mathrm{d}z = \{2\pi g'(z)\}^{-1/2}\,z^{-E-1}\,Z^q Z_n, \qquad (5.6.12)$$

where $g(z)$ and z are, as usual, fixed by the equation

$$g(z) \equiv -\frac{E}{z} + \frac{\mathrm{d}}{\mathrm{d}z}\ln(Z^q Z_n) = 0. \qquad (5.6.13)$$

Now

$$\ln(Z^q Z_n) = q\ln Z + \ln Z_n. \qquad (5.6.14)$$

Since q is being taken very large, we shall ignore the term $\ln Z_n$ by comparison with $q\ln Z$. Further, since R is very large by comparison with M and q can only take values in the range $R - M \leqslant q \leqslant R$, the ratio E/q will never differ appreciably from E/R. Thus, we shall take $g(z)$ and z to be determined by the equation

$$g(z) \equiv -\frac{E}{z} + R\frac{\mathrm{d}}{\mathrm{d}z}\ln Z = 0. \qquad (5.6.15)$$

These approximations have eliminated the dependence of $g(z)$ and z on n and permit us to deduce from equation (5.6.11) that

$$G = \{2\pi g'(z)\}^{-1/2}z^{-E-1}\sum_n Z^q Z_n = \{2\pi g'(z)\}^{-1/2}z^{-E-1}X. \qquad (5.6.16)$$

The next step is to calculate the number of members of the Gibbs ensemble for which S has n particles of type A and is in the energy eigenstate e_{kn}. This is

$$\sum \frac{q!}{a_1! a_2! \dots}, \qquad (5.6.17)$$

where $q = R - n$ and the sum is taken over all sets $\{a_i\}$ satisfying the constraints

$$\sum_i a_i = q, \qquad \sum_i a_i \varepsilon_i = E - e_{kn}. \qquad (5.6.18)$$

A familiar argument leads to the contour integral

$$\frac{1}{2\pi\iota}\oint z^{-E+e_{kn}-1}\,Z^q\mathrm{d}z. \qquad (5.6.19)$$

for this number and, thence, to the approximation

$$\{2\pi g'(z)\}^{-1/2}z^{-E+e_{kn}-1}\,Z^q, \qquad (5.6.20)$$

where $g(z)$ and z are again given by equation (5.6.15) to a high degree of accuracy. Dividing this result by G (given by equation (5.6.16)), we find for the probability S has n type A particles and is in the state e_{kn},

$$p_{kn} = z^{e_{kn}}Z^q/X = z^{e_{kn}}Z^{-n}/\mathcal{Z}, \qquad (5.6.21)$$

where

$$\mathscr{Z} = Z^{-R} X = Z^{-R} \sum_n Z^q Z_n = \sum_n Z^{-n} Z_n, \qquad (5.6.22)$$

having used equation (5.6.4). $\mathscr{Z}$ is called the *grand partition function* for the system S.

Evidently, had we chosen to specify the system's environment differently, this would only have altered the form of Z—the main characteristics of the last result would have remained unchanged. Putting $Z = e^{-\mu\beta}$, therefore, we expect the parameter μ to characterize the environment sufficiently for our purpose; it is called the *chemical potential*. Then, also putting $z = e^{-\beta}$, we can express our last result in the form

$$p_{kn} = \exp\{\beta(\mu n - e_{kn})\} / \mathscr{Z} = -\frac{1}{\beta} \frac{\partial}{\partial e_{kn}} (\ln \mathscr{Z}), \qquad (5.6.23)$$

where

$$\mathscr{Z} = \sum_{n,\,k} \exp\{\beta(\mu n - e_{kn})\} = \sum_n e^{\beta\mu n} Z_n. \qquad (5.6.24)$$

Note that the grand partition function is a weighted sum of the petit partition functions of S with $n = 0, 1, 2, \ldots$ particles of type A.

We can now calculate the mean energy of S and the mean number of type A particles it contains:

Thus, the mean number of type A particles is

$$N = \bar{n} = \sum_{k,\,n} n p_{kn} = \mathscr{Z}^{-1} \sum n \exp\{\beta(\mu n - e_{kn})\} = \frac{1}{\beta\mathscr{Z}} \frac{\partial\mathscr{Z}}{\partial\mu} = \frac{1}{\beta} \frac{\partial}{\partial\mu} (\ln \mathscr{Z}),$$
$$(5.6.25)$$

and the mean energy of S is

$$U = \bar{e} = \sum_{k,\,n} p_{kn} e_{kn} = \mathscr{Z}^{-1} \sum e_{kn} \exp\{\beta(\mu n - e_{kn})\}$$

$$= \mathscr{Z}^{-1} \sum (e_{kn} - \mu n) \exp\{\beta(\mu n - e_{kn})\} + \mu \mathscr{Z}^{-1} \sum n \exp\{\beta(\mu n - e_{kn})\}$$

$$= -\frac{1}{\mathscr{Z}} \frac{\partial\mathscr{Z}}{\partial\beta} + \mu N = -\frac{\partial}{\partial\beta} (\ln \mathscr{Z}) + \mu N. \qquad (5.6.26)$$

If N and U are regarded as given quantities, then the last two equations determine the values of the parameters β and μ. A grand canonical ensemble of C replicas of S can then be imagined constructed in which the number of members in the energy state e_{kn} with n type A particles is taken to be $C p_{kn}$; this is the appropriate statistical model for a system free to exchange heat and particles with its environment.

5.7 Thermodynamics of an open system

Suppose the environment of each of the systems in the grand canonical ensemble is varied infinitesimally in the same way —thus, both the temperature and the concentration of type A particles in these surroundings can change. At the same time, let the external fields also change infinitesimally so that work is performed on each system. Then the mean over the ensemble for a system's change of energy is

$$dU = \sum p_{kn} de_{kn} + \sum e_{kn} dp_{kn}. \tag{5.7.1}$$

The first term in this expression for dU gives the average work done by the external fields and the second term gives the average energy increase due to the heat and type A particles transferred to the systems from the bath. Thus, if dQ is the average energy transferred as heat and dA is the average energy transferred as particles, then

$$dQ = dU - \sum p_{kn} de_{kn} - dA. \tag{5.7.2}$$

Hence

$$\beta \, dQ = d(\beta U) - U \, d\beta + \sum \frac{\partial}{\partial e_{kn}} (\ln \mathscr{Z}) \, de_{kn} - \beta \, dA, \tag{5.7.3}$$

by use of equation (5.6.23). Since $\mathscr{Z} = \mathscr{Z}(e_{kn}, \beta, \mu)$, we have

$$d(\ln \mathscr{Z}) = \sum \frac{\partial}{\partial e_{kn}} (\ln \mathscr{Z}) de_{kn} + \frac{\partial}{\partial \beta} (\ln \mathscr{Z}) d\beta + \frac{\partial}{\partial \mu} (\ln \mathscr{Z}) d\mu. \tag{5.7.4}$$

This equation, together with equations (5.6.25) and (5.6.26), permit our writing equation (5.7.3) in the form

$$\beta \, dQ = d(\beta U) + d(\ln \mathscr{Z}) - \mu N \, d\beta - \beta N \, d\mu - \beta \, dA. \tag{5.7.5}$$

If $\beta = 1/kT$, dQ is known to be an exact differential, viz. $(1/k)dS$, where S is the system's entropy. But the right-hand member of equation (5.7.5) is such a differential only provided $dA = \mu dN$—then

$$\beta \, dQ = dS/k = d(\beta U + \ln \mathscr{Z} - \mu \beta N). \tag{5.7.6}$$

Thus, thermodynamical considerations require that

$$\beta = 1/kT, \qquad S = k \ln \mathscr{Z} + (U - \mu N)/T, \tag{5.7.7}$$

$$dA = \mu dN. \tag{5.7.8}$$

Equations (5.7.7) should be compared with equations (4.3.23) for a closed system.

Assuming that, for the type of system usually encountered, the departure of observed values of macroscopic variables from their averages over the grand canonical ensemble are negligible, we can accept these equations as valid for measurements made on the system. The energy equation for an infinitesimal quasi-static change in the system now follows from equation (5.7.2) in the form

$$dU = T \, dS - \sum F_r \, d\theta_r + \mu dN, \tag{5.7.9}$$

where F_r is a generalized component of force exerted *by the system* on the external fields. Evidently, μ is the energy per particle transferred to S from the bath.

That fluctuations about the means for macroscopic variables are small can be checked by computation of their standard deviations. Thus, the second moment over the ensemble of the particle number n is

$$\sum n^2 p_{kn} = \mathscr{Z}^{-1} \sum n^2 \exp\{\beta(\mu n - e_{kn})\} = \frac{1}{\beta^2 \mathscr{Z}} \frac{\partial^2 \mathscr{Z}}{\partial \mu^2}. \qquad (5.7.10)$$

It now follows that the variance of n is given by

$$\mathrm{var}(n) = \frac{1}{\beta^2 \mathscr{Z}} \frac{\partial^2 \mathscr{Z}}{\partial \mu^2} - \bar{n}^2 = \frac{1}{\beta^2} \frac{\partial^2}{\partial \mu^2} (\ln \mathscr{Z}), \qquad (5.7.11)$$

having used equation (5.6.25). Hence, if Δn is the standard deviation,

$$\Delta n/\bar{n} = \sqrt{\frac{\partial^2}{\partial \mu^2}(\ln \mathscr{Z})} \bigg/ \frac{\partial}{\partial \mu}(\ln \mathscr{Z}). \qquad (5.7.12)$$

Now, for a system comprising a large number M of identical subsystems, the quantities $S, \ln \mathscr{Z}, U, N$ occurring in equation (5.7.7) are all proportional to M (these are termed *extensive quantities*, to distinguish them from other *intensive variables* like pressure and temperature, which are unaffected when the system is enlarged without other change of state). Thus, $\Delta n/\bar{n} \propto 1/\sqrt{M}$ and is therefore very small for values of $M(\sim 10^{23})$ commonly to be expected when the subsystems are molecules.

To calculate the generalized components of force F_r, we note that

$$\sum F_r d\theta_r = -\sum p_{kn} de_{kn} = \frac{1}{\beta} \sum \frac{\partial}{\partial e_{kn}} (\ln \mathscr{Z}) de_{kn}. \qquad (5.7.13)$$

But $\mathscr{Z} = \mathscr{Z}(\theta_r, \beta, \mu) = \mathscr{Z}(e_{kn}, \beta, \mu)$ and it follows that

$$\sum \frac{\partial}{\partial \theta_r} (\ln \mathscr{Z}) d\theta_r = \sum \frac{\partial}{\partial e_{kn}} (\ln \mathscr{Z}) de_{kn}. \qquad (5.7.14)$$

(See derivation of equation (3.6.10).) Thus,

$$\sum F_r d\theta_r = \frac{1}{\beta} \sum \frac{\partial}{\partial \theta_r} (\ln \mathscr{Z}) d\theta_r, \qquad (5.7.15)$$

which leads to the result

$$F_r = \frac{1}{\beta} \frac{\partial}{\partial \theta_r} (\ln \mathscr{Z}). \qquad (5.7.16)$$

5.8 Classical gas of indistinguishable molecules

As an example of the use of the grand partition function, we now return to a consideration of the classical Maxwell gas of non-interacting molecules first studied in sections 3.5–3.7. Imagine a closed surface to be drawn within the container holding the gas, and suppose the surface encloses a volume V of the gas and N molecules. This specimen of the gas may be thought of as immersed in the heat and particle bath provided by the rest of the gas. Then, by calculating the grand partition function for this specimen and using results established in the last section, we shall be able to calculate its energy, pressure and entropy.

Denoting the partition function for a single molecule confined to the volume V by Z, we must first calculate the partition function Z_n for n such molecules. If ε_i is the energy for a molecule in the ith cell of the molecular phase space, then a state of the gas for which a_1 molecules belong to cell 1, a_2 molecules to cell 2, and so on, will have energy

$$e = a_1\varepsilon_1 + a_2\varepsilon_2 + \ldots . \tag{5.8.1}$$

This time, we shall take account of the indistinguishability of the molecules and thus, according to the principles of quantum mechanics, this state of the gas will be counted once only; i.e. the states obtained by exchange of gas molecules between cells will not be counted as additional distinguishable states. In section 3.5, this state was referred to as a condition and corresponded to W (see equation (3.5.1)) separate states of the microcanonical ensemble, all equally probable. Thus, the gas partition function Z_n is now given by

$$Z_n = \sum z^{a_1\varepsilon_1 + a_2\varepsilon_2 + \cdots} \tag{5.8.2}$$

summed over all sets of integers a_i satisfying the constraint

$$\sum a_i = n. \tag{5.8.3}$$

Note that Z_n is no longer equal to Z^n as it is in the pure classical theory. In fact,

$$Z^n = (z^{\varepsilon_1} + z^{\varepsilon_2} + \ldots)^n = \sum \frac{n!}{a_1!a_2!\ldots} z^{a_1\varepsilon_1 + a_2\varepsilon_2 + \cdots} \tag{5.8.4}$$

again summed over all sets of integers $\{a_i\}$ satisfying the constraint (5.8.3). But, suppose we choose the size of the cells in the molecular phase space to be so small that, in the very large majority of states, very few are occupied and the a_i take values 0 or 1 only. Then, for most states $a_i! = 1$ and, to a high degree of accuracy,

$$Z^n = n!\sum z^{a_1\varepsilon_1 + a_2\varepsilon_2 + \cdots} = n!Z_n. \tag{5.8.5}$$

Thus, we shall take

$$Z_n = Z^n/n!. \tag{5.8.6}$$

Substitution from the last equation into equation (5.6.24) now yields for the grand partition function the result

$$\mathscr{Z} = \sum_{n=0}^{\infty} e^{\beta\mu n} Z^n/n! = \exp(e^{\beta\mu}Z). \tag{5.8.7}$$

Equation (5.6.25) now gives the mean number of gas molecules in the volume V, viz.

$$N = e^{\beta\mu} Z \qquad \text{or} \qquad \mu = \frac{1}{\beta}\ln(N/Z). \qquad (5.8.8)$$

Regarding N as given, this equation determines μ.
The energy of the gas is calculated from equation (5.6.26) to be

$$U = -e^{\beta\mu} \partial Z/\partial\beta = -N\frac{\partial}{\partial\beta}(\ln Z). \qquad (5.8.9)$$

This is identical with the first of equations (3.5.15).
The gas pressure follows from equation (5.7.6), thus:

$$P = \frac{1}{\beta}\frac{\partial}{\partial V}(\ln \mathscr{Z}) = kNT\frac{\partial}{\partial V}(\ln Z), \qquad (5.8.10)$$

which is equation (3.6.13) again.
Finally, we calculate S from equation (5.7.7) to be given by

$$S = kN\left[\ln Z - \beta\frac{\partial}{\partial\beta}(\ln Z) + 1 - \ln N\right]. \qquad (5.8.11)$$

This is no longer in agreement with equation (3.6.6), the last two terms being absent from the latter equation. But, as pointed out in section 3.7, it is necessary to subtract a term $kN\ln N$ from the earlier formula to ensure that $S \propto N$ and our revised analysis happily justifies this amendment. Thus, using the partition function (3.7.9) for a point molecule, we now calculate that

$$S = kN\left[\frac{5}{2}\ln T - \ln P + \frac{5}{2} + \ln k + \frac{3}{2}\ln(2\pi mk/h^2)\right]. \qquad (5.8.12)$$

(Cf. equation (3.7.14).)

Exercises 5

1. In the special case where all the systems of an ensemble are in the same pure state ψ, show that the density matrix is a projection operator, i.e. $\rho^2 = \rho$. Deduce that ρ then has eigenvalues 1 and 0, the corresponding eigenstates being ψ and all states orthogonal to ψ, respectively.

2. Prove the identity

$$\text{Tr}(\rho^2) = \sum_{r,s} P_r P_s |(\psi_r, \psi_s)|^2.$$

Deduce that, if ρ is a projection operator, then

$$\sum_{r,s} P_r P_s \{1 - |(\psi_r, \psi_s)|^2\} = 0$$

and hence that $|(\psi_r, \psi_s)| = 1$ for all r and s. Now deduce the converse of the previous exercise.

3. A certain pure state of a system is specified by the normalized vector ϕ. An ensemble of such systems is described by the density matrix ρ. Prove that

$$\text{Tr}(\rho \phi \phi^\dagger) = \sum_r P_r |(\psi_r, \phi)|^2.$$

Deduce that, if a member of the ensemble is chosen at random, the probability it will be measured to be in the state ϕ is $\text{Tr}\,(\rho \phi \phi^\dagger)$.

4. If ρ is the density matrix for a certain ensemble, show that ρ^2 satisfies the equation

$$i\hbar \, \frac{\text{d}}{\text{d}t} (\rho^2) = [H, \rho^2].$$

Deduce that, if all the systems of the ensemble are in the same pure state at time $t = t_0$, then the systems will also be in identical states at any later time $t > t_0$. (Hint: Use the results of exercises 1 and 2 above.)

5. Show that the variance of the energy calculated over the grand canonical ensemble has value

$$(\Delta U)^2 = \frac{\mu}{\beta} \frac{\partial U}{\partial \mu} - \frac{\partial U}{\partial \beta}.$$

For a system comprising a large number M of subsystems, assuming $U \propto M$ show that the relative standard deviation of the energy $\Delta U / U \propto 1/\sqrt{M}$.

CHAPTER 6

Crystals and magnets

6.1 Einstein's model for a crystal

We shall first illustrate the general theory developed in the previous chapter by applying it to a system of atoms arranged according to a uniform pattern in a crystal lattice. The atoms' motions are restricted to vibrations about positions of equilibrium located at their lattice sites and it is therefore meaningful to distinguish the atoms by their locations, i.e. the crystal state in which atom 1 is at site A and atom 2 is at site B is quite different from the state in which atom 1 is at B and atom 2 is at A; exchange of the atoms could, in principle, be effected by a prescribable physical procedure. Thus, the principle of indistinguishability of atoms, which plays an important role in the theory of gases (Chapter 7), makes no contribution to our present analysis. However, unlike the case of a dilute gas, neighbouring crystal atoms interact powerfully and it is not permissible, therefore, to ignore these interactions.

One way of meeting this difficulty is to approximate the situation by replacing the actual forces applied by its neighbours to an atom, by an average field of force in which the atom is assumed to move independently. The atoms in such a crystal model are then treated as independent subsystems and, if Z is the partition function for an individual atom, the partition function for the crystal is Z^N, where N is the number of atoms. In this section, we shall analyse such a quantum mechanical model for a crystal, which was proposed by Einstein in 1907.

This treats each atom (mass m) as an independent three-dimensional harmonic oscillator, attracted towards its equilibrium position by a force $m\omega^2 r$ proportional to the distance r from this position. Such an oscillator has energy eigenvalues $\hbar\omega(n+3/2) = h\nu(n+3/2)$ ($\omega = 2\pi\nu$, $n = 0, 1, 2, \ldots$), the nth level being $\frac{1}{2}(n+1)$ $(n+2)$-fold degenerate (see e.g. D. F. Lawden, *Mathematical Principles of Quantum Mechanics*, Methuen, 1967, p. 260). Its partition function is accordingly calculated to be

$$\sum_{n=0}^{\infty} \frac{1}{2}(n+1)(n+2)\exp\left\{-\beta h\nu(n+3/2)\right\}$$

$$= \alpha^{3/2} \sum_{n=0}^{\infty} \frac{1}{2}(n+1)(n+2)\alpha^n = \alpha^{3/2}(1-\alpha)^{-3}, \qquad (6.1.1)$$

where $\alpha = \exp(-\beta h\nu) = \exp(-h\nu/kT)$. The assumption being that the atomic motions are independent, the partition function for the whole crystal of N atoms will be Z, where

$$\ln Z = N \ln \{\alpha^{3/2}(1-\alpha)^{-3}\} = -\frac{3}{2} N\beta h\nu - 3N \ln (1 - e^{-\beta h\nu}). \quad (6.1.2)$$

The mean energy of the crystal now follows from equation (4.3.16) and can be expressed in the form

$$U = \frac{3}{2} Nh\nu \coth (h\nu/2kT). \quad (6.1.3)$$

Thus, the heat capacity is given by

$$C_V = \partial U/\partial T = \frac{3Nh^2\nu^2}{4kT^2} \operatorname{cosech}^2 (h\nu/2kT)$$

$$= 3Nk \left(\frac{\theta_E}{T}\right)^2 \frac{e^{\theta_E/T}}{(e^{\theta_E/T}-1)^2}, \quad (6.1.4)$$

where $\theta_E = h\nu/k$ is called the *Einstein characteristic temperature*.

If T is large compared with θ_E, so that θ_E/T is small, it is easy to approximate equation (6.1.4) by

$$C_V = 3Nk \left\{1 - \frac{1}{12} (\theta_E/T)^2 + \dots \right\}. \quad (6.1.5)$$

Clearly, as $T \to +\infty$, the heat capacity approaches the Dulong–Petit value (section 4.7) from below. If T is small compared with θ_E, so that θ_E/T is large, a good approximation is

$$C_V = 3Nk(\theta_E/T)^2 e^{-\theta_E/T} \quad (6.1.6)$$

and thus, at $T \to 0$, $C_V \to 0$. This accounts for some crystals having atomic heats well below the Dulong–Petit value of 6 at room temperature, viz. their characteristic temperatures are much higher than normal room temperature. Raising the temperature of such crystals would be expected to cause their atomic heats to increase towards the value 6. Thus, the atomic heat of diamond at 300 K is about 1.35, but at 1400 K this has risen to 5.5, in conformity with the Einstein model; the characteristic temperature for diamond is 1860 K.

6.2   Debye's model for a crystal. Phonons

A crystal model, more closely in agreement with experiment than Einstein's, was constructed by P. J. Debye (1884–1966) in 1912. Instead of treating the crystal atoms as independent vibrators, Debye's idea was to base the analysis on the normal modes of vibration of the crystal as a unified system. Since $3N$ coordinates are needed to specify the positions of the N atoms, we expect there to be $3N$ normal modes of oscillation of the complete crystal, the low frequency modes corresponding to macroscopic elastic vibrations of the solid and the modes of very high frequency corresponding to microscopic vibrations of the individual atoms. More precisely, six of the $3N$ generalized coordinates will be used to specify the position and orientation of the body of the crystal in space, and only the remaining $3N-6$ coordinates will contribute to the normal coordinates whose separate oscillations determine the normal modes. However, N is always so large that this discrepancy can be ignored.

If, then, q_i is a normal coordinate for the crystal, whose simple harmonic oscillation corresponds to the ith normal mode, classical mechanics requires that it satisfies an equation of motion

$$\ddot{q}_i + (2\pi v_i)^2 q_i = 0, \tag{6.2.1}$$

where v_i is the frequency of the resulting oscillation. According to the principles of quantum mechanics, a particle whose motion is governed by such a classical equation constitutes a harmonic oscillator and its energy is quantized with eigenvalues $\varepsilon_n = (n+\tfrac{1}{2})hv_i$ $(n = 0, 1, 2, \ldots)$. Debye treated each normal mode as just such a harmonic oscillator and regarded all these harmonic oscillators as virtually independent systems. Thus, the Debye-model represents a crystal as an ideal gas of harmonic oscillators.

As explained earlier, v will have a discrete spectrum of $3N$ eigenvalues, but N is so large that it is more convenient to treat the spectrum as being continuous. We define a spectrum density function $g(v)$, such that $g(v)dv$ is the number of normal modes in the frequency range $(v, v + dv)$. If v_m denotes the largest frequency in the spectrum, we must have

$$\int_0^{v_m} g(v)dv = 3N. \tag{6.2.2}$$

$g(v)$ will be taken to be determined by the structure of the crystal and, hence, to be independent of its temperature. It could, however, be dependent on the crystal's state of stress and we shall assume, therefore, that the crystal is not subject to external forces.

Consider the oscillator whose natural frequency is v. Its partition function is

$$\sum_{n=0}^{\infty} \exp\{-\beta(n+\tfrac{1}{2})hv\} = e^{-\frac{1}{2}\beta hv}/(1 - e^{-\beta hv}). \tag{6.2.3}$$

Since these normal modes are to be treated as independent subsystems, the partition function Z for the crystal is the product of the partition functions for the normal modes. It follows that

$$\ln Z = -\int_0^{v_m} g(v)\left[\tfrac{1}{2}\beta hv + \ln(1 - e^{-\beta hv})\right]dv. \tag{6.2.4}$$

The mean energy of the crystal can now be found by differentiating the last equation with respect to β, to give

$$U = -\frac{\partial}{\partial \beta}(\ln Z)$$

$$= \int_0^{v_m} g(v)\left(\frac{1}{2}hv + \frac{hv}{e^{\beta hv} - 1}\right)dv = U_0 + \int_0^{v_m}\frac{g(v)hv}{e^{\beta hv} - 1}dv, \qquad (6.2.5)$$

where

$$U_0 = \frac{1}{2}\int_0^{v_m} g(v)hv\,dv \qquad (6.2.6)$$

is independent of β. As $T \to 0$, $\beta \to +\infty$ and $U \to U_0$. Thus U_0 is the internal energy of the crystal at absolute zero temperature—the so-called *zero-point energy*. Taking the zero-point energy as the datum from which we measure the internal energy, we can put $U_0 = 0$ and so simplify the energy equation to the form

$$U = \int_0^{v_m}\frac{g(v)hv}{e^{\beta hv} - 1}dv. \qquad (6.2.7)$$

Further differentiation yields the heat capacity, viz.

$$C_V = \partial U/\partial T = -k\beta^2\frac{\partial U}{\partial \beta} = \int_0^{v_m} g(v)\frac{k(\beta hv)^2 e^{\beta hv}}{(e^{\beta hv} - 1)^2}dv. \qquad (6.2.8)$$

In the case of the electromagnetic field, the energy of each mode of oscillation is quantized in units of hv, where v is the frequency of the mode. Thus, electromagnetic radiation can be modelled as a photon gas, each photon having energy ε and frequency v related by the Einstein equation $\varepsilon = hv$. This model has been employed in section 2.6 and will be analysed from the standpoint of statistical mechanics in section 7.6. It will there be shown (equation (7.6.8)) that the mean number of photons with energy ε is $1/(e^{\beta \varepsilon} - 1)$ and, if V is the volume of the enclosure containing the radiation, the density $g(v)$ of the normal modes having frequency v is $8\pi V v^2/c^3$ (equation (7.6.9)).

This model can be adapted to our present purpose. Thus, suppose we introduce a hypothetical particle called a *phonon*, whose energy is related to the frequency v of the normal mode of vibration of the crystal with which it is associated by the Einstein equation $\varepsilon = hv$. Then, if the crystal is modelled as a gas of non-interacting phonons, the number having energy ε being taken to be $1/(e^{\beta \varepsilon} - 1)$, the total energy of the crystal will be given by equation (6.2.7). Thus, the model of the crystal as a phonon gas is consistent with our analysis.

Debye assumed $g(v) \propto v^2$ as for photons. Thus, putting $g(v) = \kappa v^2$ in equation (6.2.2), we find $\kappa = 9N/v_m^3$ and, hence,

$$g(v) = 9Nv^2/v_m^3. \qquad (6.2.9)$$

Substitution in equation (6.2.8) now leads to the result

$$C_V = 9Nk(T/\theta_D)^3 \int_0^{\theta_D/T} \frac{x^4 e^x}{(e^x - 1)^2} \, dx, \tag{6.2.10}$$

where $\theta_D = h\nu_m/k$ is called the *Debye characteristic temperature.*

If T is large compared with θ_D, x ranges over small values only and the integrand can be approximated by a power series to give

$$C_V = 9Nk(T/\theta_D)^3 \int_0^{\theta_D/T} x^2 (1 - x^2/12 + \ldots) \, dx,$$

$$= 3Nk\{1 - (\theta_D/T)^2/20 + \ldots\}, \tag{6.2.11}$$

showing that as $T \to \infty$, C_V approaches the Dulong–Petit value of $3Nk$ from below.

If T is small compared with θ_D, the upper limit of integration can be approximated by $+\infty$ and the integral then computed thus:

$$\int_0^\infty x^4 e^{-x} (1 - e^{-x})^{-2} \, dx = \int_0^\infty x^4 \sum_{n=1}^\infty n e^{-nx} \, dx$$

$$= 24 \sum_{n=1}^\infty n^{-4} = 4\pi^4/15. \tag{6.2.12}$$

Hence,

$$C_V = \frac{12}{5} \pi^4 Nk(T/\theta_D)^3, \tag{6.2.13}$$

i.e. $C_V \propto T^3$ at low temperatures, a prediction which is supported by the experimental data.

It is possible to calculate ν_m, and therefore θ_D, from the elastic constants of the crystal. Thus, if we assume for a particular type of elastic vibration arising from waves having velocity of propagation c, that the spectrum density function $g(\nu)$ is given by the same formula as for electromagnetic radiation, then

$$g(\nu) = \frac{4\pi V}{c^3} \nu^2. \tag{6.2.14}$$

(Note: A factor 2, which is introduced to allow for the two modes of polarization of electromagnetic waves, has been omitted; the appropriate factor for the distinct elastic modes will be given immediately.) Two types of wave are possible in an elastic solid, longitudinal (or compressional) and transverse (or shear) (see J. A. Hudson, *The Excitation and Propagation of Elastic Waves*, C.U.P., 1980) and the transverse waves are separable into two distinct components. If c_L, c_T are the velocities of propagation of longitudinal and transverse waves respectively, then

$$c_L^2 = (\lambda + 2\mu)/\rho, \qquad c_T^2 = \mu/\rho, \tag{6.2.15}$$

where λ, μ are the Lamé elastic constants and ρ is the material's density. Debye assumed

$$g(v) = 4\pi V \left(\frac{1}{c_L^3} + \frac{2}{c_T^3} \right) v^2, \qquad (6.2.16)$$

where both c_L and c_T need averaging over the various possible directions of propagation in a crystal (note the factor 2 associated with the two components of a transverse wave). Equation (6.2.2) now gives

$$v_m^2 = \frac{9N}{4\pi V} \left(\frac{1}{c_L^3} + \frac{2}{c_T^3} \right)^{-1} \qquad (6.2.17)$$

and $\theta_D = h v_m / k$ now yields the characteristic temperature. This formula for θ_D gives values in good agreement with calorimetric data.

6.3   Quantum theory of paramagnetic materials

In this section, we shall re-examine the model for a paramagnetic material already analysed using classical statistics in section 4.6, this time from the standpoint of quantum theory. It should be recalled that each atom of the material is taken to be a magnetic dipole which pivots freely about its centre under the influence of an external field, but that it is assumed the magnetic interactions between the molecules are negligible.

According to quantum theory, the magnetic dipole moment **m** of a molecule is related to its overall angular momentum **J** (i.e. the resultant of the orbital angular momenta of its electrons and the spins of its constituent particles, measured in units of $\hbar$) by the Landé formula

$$\mathbf{m} = g\mu_B \mathbf{J}, \qquad (6.3.1)$$

where $\mu_B = e\hbar / 2m_e$ is the *Bohr magneton* (e = electronic charge; m_e = electronic mass) and g is the Landé-factor. Thus, the potential energy of the dipole in the external field **H** is

$$\varepsilon = -\mu_0 \mathbf{m} \cdot \mathbf{H} = -\mu_0 g\mu_B H J_z, \qquad (6.3.2)$$

the z-axis being taken parallel to the field.

To this should now be added the rotational kinetic energy of the molecule (modelled as a rotator) and then the eigenvalues of the total energy observable calculated for use in constructing the partition function. We shall, however, avoid this complicated calculation by neglecting the rotational energy, pleading in justification that, at low temperatures, where the divergence between the quantum and classical results is most pronounced, the contribution of the kinetic energy is negligible.

If this approximation is accepted, the eigenvalues of the energy ε given by equation (6.3.2) follow immediately from the known eigenvalues of the angular momentum component J_z. For a total angular momentum quantum number J (always a positive integral multiple of $\frac{1}{2}$; $J = 0$ is a possible eigenvalue, but yields zero moment) the eigenvalues of J_z are $(k - J)$ $(k = 0, 1, 2, \ldots, 2J)$. Thus, the eigenvalues of ε are

$$\varepsilon_k = -\mu_0 g \mu_B H (k - J) = \mu_0 m H (1 - k/J), \tag{6.3.3}$$

where $m = g\mu_B J$ is the maximum value of m_z (usually regarded as the magnitude of **m**). This formula is the quantum replacement for the classical formula $\varepsilon = \mu_0 m H \cos \theta$.

The partition function for a molecule can now be calculated by the summation of a geometric series, thus:

$$Z = \sum_{k=0}^{2J} \exp(-\beta \varepsilon_k) = \sum_{k=0}^{2J} \exp\{\mu_0 m H \beta(k/J - 1)\}$$

$$= \frac{\sinh\{\alpha(1 + 1/2J)\}}{\sinh(\alpha/2J)}, \tag{6.3.4}$$

where $\alpha = \beta \mu_0 m H = \mu_0 m H / kT$ (as in section 4.6).

The probability of finding the molecule in the energy state ε_k now follows from equation (4.3.15) as

$$P_k = \frac{\sinh(\alpha/2J)}{\sinh\{\alpha(1 + 1/2J)\}} \exp\{\alpha(k/J - 1)\}. \tag{6.3.5}$$

In this state, $m_z = g\mu_B J_z = m(k/J - 1)$ and the mean (or expected) value of m_z is accordingly

$$\bar{m}_z = \sum_{k=0}^{2J} P_k m_z(k) = m \frac{\sinh(\alpha/2J)}{\sinh\{\alpha(1 + 1/2J)\}} \sum_{k=0}^{2J} (k/J - 1)\exp\{\alpha(k/J - 1)\}$$

$$= m \frac{\sinh(\alpha/2J)}{\sinh\{\alpha(1 + 1/2J)\}} \frac{\partial}{\partial \alpha} \sum_{k=0}^{2J} \exp\{\alpha(k/J - 1)\}$$

$$= m \frac{\sinh(\alpha/2J)}{\sinh\{\alpha(1 + 1/2J)\}} \frac{\partial}{\partial \alpha} \frac{\sinh\{\alpha(1 + 1/2J)\}}{\sinh(\alpha/2J)}$$

$$= m [(1 + 1/2J)\coth\{\alpha(1 + 1/2J)\} - (1/2J)\coth(\alpha/2J)].$$

$$\tag{6.3.6}$$

If the paramagnetic specimen has N molecules per unit volume, its intensity of magnetization now follows as

$$I = Nm[(1 + 1/2J)\coth\{\alpha(1 + 1/2J)\} - (1/2J)\coth(\alpha/2J)]. \tag{6.3.7}$$

The formula just obtained should be compared with the classical result (4.6.8). As $J \to \infty$, $(1/2J)\coth(\alpha/2J) \to 1/\alpha$ and the two formulae converge (as we expect, since the number of energy eigenstates for a molecule then becomes large and the distinction between the discrete and continuous spectra of the two theories vanishes). For large values of α (low temperatures and intense fields), we obtain, as in the classical theory, $I = I_\infty = Nm$, i.e. the material reaches magnetic saturation. For small values of α (higher temperatures and weak fields), we calculate that

$$I = \frac{1}{3} N m \alpha (1 + 1/J) = \mu_0 N m^2 H (1 + 1/J)/3kT \qquad (6.3.8)$$

and the susceptibility factor is given by

$$\chi = \mu_0 N m^2 (1 + 1/J)/3kT. \qquad (6.3.9)$$

We note that the Curie constant is $\mu_0 N m^2 (1 + 1/J)/3k$, as compared with the classical value of $\mu_0 N m^2 /3k$. Thus, for $J = \frac{1}{2}$, 1, this constant takes values $\mu_0 N m^2 /k$, $2\mu_0 N m^2 /3k$, respectively.

The formula (6.3.7) has been checked experimentally by observation of the paramagnetic characteristics of the rare earth elements, for which the angular momentum quantum numbers J have a wide range. Good agreement is found.

Exercises 6

1. Show that the relative standard deviation of the energy of an Einstein crystal is

$$\frac{1}{\sqrt{(3N)}} \operatorname{sech}(\theta_E/2T).$$

(Note: This is less than $1/\sqrt{(3N)}$ even for very high temperatures.)

2. Show that the entropy of an Einstein crystal is given by

$$S = 3Nk \left[\frac{x}{e^x - 1} - \ln(1 - e^{-x}) \right],$$

where $x = \theta_E/T$. Verify that $S \to 0$ as $T \to 0$, in accordance with the third law.

3. Integrating by parts, show that an alternative formula to equation (6.2.10) for the heat capacity of a Debye crystal is

$$C_V = 9Nk \left[\frac{4}{x_D^3} \int_0^{x_D} \frac{x^3 dx}{e^x - 1} - \frac{x_D}{e^{x_D} - 1} \right],$$

where $x_D = \theta_D/T$.

4. Show that the relative standard deviation of the energy of a Debye crystal is

$$\frac{1}{3N^{1/2}I} [4I - (\theta_D/T)/\{\exp(\theta_D/T) - 1\}]^{1/2}$$

where

$$I = (T/\theta_D)^3 \int_0^{\theta_D/T} \frac{x^3 dx}{e^x - 1}.$$

If T is large compared with θ_D, show that this result reduces to

$$\frac{1}{\sqrt{(3N)}}\left[1+\frac{3}{8}(\theta_D/T)+O(\theta_D/T)^2\right].$$

5. Show that, if the zero-point energy is included, the energy of a Debye crystal is given by

$$U = 9Nk\theta_D\left[\frac{1}{8}+(T/\theta_D)^4\int_0^{\theta_D/T}\frac{x^3\,\mathrm{d}x}{e^x-1}\right]$$

and deduce, using equation (4.3.23), that its entropy is given by

$$S = 3Nk\left[4(T/\theta_D)^3\int_0^{\theta_D/T}\frac{x^3\,\mathrm{d}x}{e^x-1}-\ln\{1-\exp(-\theta_D/T)\}\right].$$

Verify that $S \to 0$ as $T \to 0$.

6. The classical formula for the energy of a certain three-dimensional particle oscillator is

$$\varepsilon = \frac{p^2}{2m}+\frac{1}{2}m(\omega r)^2+\frac{1}{6}\alpha m^2(\omega r)^4,$$

where p is the momentum and r is the distance from the equilibrium position. Taking α to be small and working to $O(\alpha)$, show that the classical partition function for the oscillator is

$$Z = \left(\frac{2\pi}{\omega\beta h}\right)^3(1-5\alpha/2\beta),$$

where $\beta = 1/kT$. Deduce that a crystal modelled by N such oscillators has heat capacity given by

$$C_V = Nk(3-5\alpha kT).$$

7. Show that the entropy of the paramagnetic material modelled in section 6.3 is given by

$$S = Nk\ln\left[\frac{\sinh\{\alpha(1+1/2J)\}}{\sinh(\alpha/2J)}\right]+Nk\alpha[(1/2J)\coth(\alpha/2J)$$
$$-(1+1/2J)\coth\{\alpha(1+1/2J)\}],$$

where $\alpha = \mu_0 mH/kT$. Verify that $S \to 0$ as $T \to 0$.

CHAPTER 7

Bose–Einstein and Fermi–Dirac statistics

7.1 Gas of indistinguishable non-interacting molecules

We now return to the problem, already treated by the classical theory in sections 3.5 and 4.4, of a gas comprising a very large number of identical molecules confined to a rigid container and whose mutual interaction is negligible. The container will be assumed immersed in a heat bath at temperature T and it, and its contents, to be in equilibrium. We shall analyse the open system comprising, on average, N molecules located inside a closed surface drawn within the container and embracing a volume V. This system can exchange heat and molecules with the rest of the gas, and will therefore be modelled by a grand canonical ensemble. Clearly, all the molecules belonging to the system are of type A, to use the terminology of section 5.6. By treating this open system, instead of the closed system provided by the whole gas, we avoid introducing a constraint, like equation (3.5.3), which expresses the conservation of the number of molecules in the system. This simplifies the mathematical analysis considerably.

Let $\varepsilon_i (i = 1, 2, \ldots)$ be the energy eigenstates for a single molecule confined to the volume V (remember that these states are also eigenstates of the other observables completing the maximal set of compatible observables and that, therefore, the energy eigenvalues will not, generally, be all different; each will be repeated a number of times equal to the order of degeneracy of the energy eigenstate). Consider a state of the gas in which a_1 molecules are in the energy eigenstate ε_1, a_2 in the eigenstate ε_2, and so on; this energy eigenstate of the gas will have energy eigenvalue

$$e_{Kn} = a_1 \varepsilon_1 + a_2 \varepsilon_2 + \ldots = \sum a_i \varepsilon_i. \tag{7.1.1}$$

where

$$\sum a_i = n, \tag{7.1.2}$$

n being the (variable) number of molecules occupying the volume V. Since the molecules are indistinguishable, exchange of molecules between the different eigenstates ε_i does not generate new states of the gas and, thus, the condition $\{a_i\}$ counts as only one state and not the large number given at (3.5.1) in the classical theory.

If the molecules are bosons, as is the case for all ordinary gases, the integers a_i

have no constraints to satisfy, and we say that the statistics are *Bose–Einstein*. If, however, the molecules are fermions (e.g. the electrons whose flow constitutes an electric current in a metal and which can be treated as a gas, see section 7.5), Pauli's exclusion principle, by which no two fermions can occupy the same state, is operative and each a_i can only take the values 0 and 1. In these circumstances, we say the statistics are *Fermi–Dirac*. In this section, we shall restrict the analysis to Bose–Einstein statistics.

Using equation (5.6.24), the grand partition function for the boson gas can now be written down, thus:

$$\mathscr{Z} = \sum z^{\Sigma a_i \varepsilon_i - \mu \Sigma a_i}, \qquad (7.1.3)$$

where $z = e^{-\beta}$. The overall summation must be carried out over all sets of positive integers (or zeros) $\{a_i\}$ and we shall assume there is no upper limit to the value of each a_i (i.e. the available number of molecules is assumed infinite). Putting

$$\zeta = z^{-\mu} = e^{\beta \mu}, \quad z_i = z^{\varepsilon_i} = e^{-\beta \varepsilon_i}, \qquad (7.1.4)$$

we calculate that

$$\mathscr{Z} = \sum \zeta^{a_1 + a_2 + \cdots} z_1^{a_1} z_2^{a_2} \ldots ,$$

$$= (1 + \zeta z_1 + \zeta^2 z_1^2 + \ldots)(1 + \zeta z_2 + \zeta^2 z_2^2 + \ldots)\ldots ,$$

$$= \prod_{i=1}^{\infty} \frac{1}{1 - \zeta z_i} \qquad (7.1.5)$$

This is the grand partition function.

Assuming the mean number of molecules in the volume V is known to be N, equation (5.6.25) now leads to the condition

$$\sum_i \frac{\zeta z_i}{1 - \zeta z_i} = N. \qquad (7.1.6)$$

This determines the value of ζ, and therefore of μ.

From equation (5.6.23), the probability of observing the gas to comprise n molecules and to have energy e_{kn} is

$$p_{kn} = \exp\{\beta(\mu n - e_{kn})\}/\mathscr{Z} = \exp\{\beta(\mu \Sigma a_i - \Sigma a_i \varepsilon_i)\}/\mathscr{Z}. \qquad (7.1.7)$$

It now follows that the mean number of molecules in the energy eigenstate ε_j is

$$\bar{a}_j = \sum a_j \exp\{\beta(\mu \Sigma a_i - \Sigma a_i \varepsilon_i)\}/\mathscr{Z} = -\frac{1}{\beta \mathscr{Z}} \frac{\partial}{\partial \varepsilon_j} \sum \exp\{\beta(\mu \Sigma a_i - \Sigma a_i \varepsilon_i)\}$$

$$= -\frac{1}{\beta \mathscr{Z}} \frac{\partial \mathscr{Z}}{\partial \varepsilon_j} = -\frac{1}{\beta} \frac{\partial}{\partial \varepsilon_j}(\ln \mathscr{Z}), \qquad (7.1.8)$$

the summations being taken over all possible sets $\{a_i\}$. Substituting for $\mathscr{Z}$ from equation (7.1.5), we find that

$$\bar{a}_j = \frac{1}{\beta} \frac{\partial}{\partial \varepsilon_j} \sum_i \ln(1 - \zeta z_i) = \frac{\zeta z_j}{1 - \zeta z_j}. \qquad (7.1.9)$$

Equation (7.1.6) can now be seen to be equivalent to the mean of equation (7.1.2), viz.

$$\sum \bar{a}_i = N. \tag{7.1.10}$$

An alternative form for equation (7.1.9) is

$$\bar{a}_j = \frac{1}{\zeta^{-1} \exp(\beta \varepsilon_j) - 1}, \tag{7.1.11}$$

which should be compared with the classical form (3.5.10) and to which it reduces if the '1' in the denominator is omitted.

The mean energy of the gas follows thus:

$$U = \sum \bar{a}_i \varepsilon_i = \sum_i \frac{\varepsilon_i}{\zeta^{-1} \exp(\beta \varepsilon_i) - 1}. \tag{7.1.12}$$

Or, this result can be derived from equation (5.6.26).

7.2 Gas of point bosons

Suppose the molecules are structureless, so that their energy is entirely due to their translatory motion. If each has mass m and they are confined to a rectangular box of dimensions $a \times b \times c$, the energy eigenvalues for a single molecule are given by

$$\varepsilon = \frac{h^2}{8m}\left(\frac{p^2}{a^2} + \frac{q^2}{b^2} + \frac{r^2}{c^2}\right), \tag{7.2.1}$$

where p, q, r, are positive, non-zero, integers. If the molecules have zero spin, a unique eigenstate is associated with each set (p, q, r), so that the energy observable alone defines a complete orthonormal set of states. If, however, each molecule has spin $n\hbar$ (where n must be a positive integer or zero for bosons), each energy eigenstate is $(2n+1)$-fold degenerate, there being $(2n+1)$ eigenstates of the spin observable s_z (z-component of spin); in these circumstances, each energy eigenstate must be repeated $(2n+1)$ times in any summation over such states. For common gases, $n = 0$ and this value will be assumed in the following calculations.

With $x = p/a$, etc., the energy eigenvalues are determined by the equation

$$\varepsilon = \frac{h^2}{8m}(x^2 + y^2 + z^2), \tag{7.2.2}$$

where the points (x, y, z) form a rectangular lattice in the positive octant of xyz-space, each cell of which has dimensions $a^{-1} \times b^{-1} \times c^{-1}$. These lattice points accordingly have density $abc = V$, the volume of the container. Now $\varepsilon = $ constant over a sphere of radius $\rho = \sqrt{(8m\varepsilon/h^2)}$ and the volume of xyz-space between the spheres $\varepsilon = $ constant, $\varepsilon + d\varepsilon = $ constant, lying in the positive octant, is therefore found to be

$$\frac{1}{2}\pi \rho^2 \, d\rho = 2\pi \left(\frac{2m}{h^2}\right)^{3/2} \varepsilon^{1/2} \, d\varepsilon. \tag{7.2.3}$$

Thus, the number of energy eigenvalues in the range $(\varepsilon, \varepsilon + d\varepsilon)$ is

$$A V \varepsilon^{1/2} d\varepsilon, \tag{7.2.4}$$

where

$$A = 2\pi (2m/h^2)^{3/2}. \tag{7.2.5}$$

We now approximate the left-hand member of equation (7.1.6) by an integral to yield the condition

$$A V \int_0^\infty \frac{\zeta e^{-\beta \varepsilon}}{1 - e^{-\beta \varepsilon}} \varepsilon^{1/2} d\varepsilon = N. \tag{7.2.6}$$

Putting $\beta \varepsilon = x$, this can be written

$$I_{1/2} = N \beta^{3/2} / A V, \tag{7.2.7}$$

where

$$I_{1/2}(\zeta) = \int_0^\infty \frac{x^{1/2} dx}{\zeta^{-1} e^x - 1}. \tag{7.2.8}$$

This condition determines ζ and μ.

Similarly, equation (7.1.12) can be approximated into the form

$$U = A V \beta^{-5/2} I_{3/2}, \tag{7.2.9}$$

where

$$I_{3/2}(\zeta) = \int_0^\infty \frac{x^{3/2} dx}{\zeta^{-1} e^x - 1}. \tag{7.2.10}$$

Further, taking logarithms of the two sides of equation (7.1.5) and approximating a sum by an integral, we find

$$\ln \mathscr{Z} = -A V \int_0^\infty \varepsilon^{1/2} \ln (1 - \zeta e^{-\beta \varepsilon}) d\varepsilon = \frac{2}{3} A V \beta^{-3/2} I_{3/2}, \tag{7.2.11}$$

after an integration by parts.

Equation (5.7.16) can now be used to derive the gas pressure, thus:

$$P = \frac{1}{\beta} \frac{\partial}{\partial V} (\ln \mathscr{Z}) = \frac{2}{3} A \beta^{-5/2} I_{3/2} = 2U/3V, \tag{7.2.12}$$

i.e. the pressure is two-thirds the energy density.

In most cases met with in practice, ζ will be small. In such circumstances, the integral $I_{1/2}$ can be expanded thus:

$$I_{1/2} = \int_0^\infty \zeta x^{1/2} e^{-x} (1 + \zeta e^{-x} + \zeta^2 e^{-2x} + \ldots) dx$$

$$= \frac{1}{2} \pi^{1/2} \sum_{j=1}^\infty \zeta^j / j^{3/2}, \tag{7.2.13}$$

having used the identity

$$\int_0^\infty s^2 e^{-\alpha s^2} ds = \frac{1}{4} \sqrt{(\pi/\alpha^3)}. \tag{7.2.14}$$

Similarly, we expand $I_{3/2}$ by the series

$$I_{3/2} = \frac{3}{4} \pi^{1/2} \sum_{j=1}^\infty \zeta^j/j^{5/2}, \tag{7.2.15}$$

using the identity

$$\int_0^\infty s^4 e^{-\alpha s^2} ds = \frac{3}{8} \sqrt{(\pi/\alpha^5)}. \tag{7.2.16}$$

If all terms after the first are ignored in these expansions, the condition (7.2.7) can first be solved to give

$$\zeta = 2N\beta^{3/2}/(\pi^{1/2} AV) = \frac{Nh^3}{V}(2\pi m k T)^{-3/2}$$

$$= 5.4 \times 10^{-27} \mu (m^* T)^{-3/2}, \tag{7.2.17}$$

where $\mu = N/V$ is the molecule density (per cubic metre) and m^* is the molecular weight. For hydrogen (H_2) at $T = 1$ K and normal pressure, $\mu = 2.7 \times 10^{25}$, $m^* = 2$, and this quantity has value about 0.052. At higher temperatures and for heavier molecules, the value will be lower and, thus, our assumption that ζ is small is amply justified.

With this value for ζ, an approximation for U follows from equation (7.2.9), viz.

$$U = \frac{3}{2} kNT, \tag{7.2.18}$$

which is the classical result (equation (4.4.4)). Equation (7.2.12) now reduces to the state equation for an ideal gas.

Again to the first order in ζ, equation (7.1.11) gives

$$\bar{a}_i = \zeta \exp(-\beta \varepsilon_i). \tag{7.2.19}$$

Now, in the eigenstate (7.2.1), p_x^2, p_y^2, p_z^2, are sharp with eigenvalues $p^2 h^2/4a^2$, $q^2 h^2/4b^2$, $r^2 h^2/4c^2$, respectively; thus, $|p_x| = ph/2a$, etc. and the number of molecules with momentum components having magnitudes $|p_x|$, $|p_y|$, $|p_z|$ is therefore

$$\zeta \exp\left(-\frac{\beta}{2m}(p_x^2 + p_y^2 + p_z^2)\right). \tag{7.2.20}$$

Accepting the classical view that these momenta are distributed equally between positive and negative values, the number of molecules with momentum components (p_x, p_y, p_z) will be

$$\frac{1}{8} \zeta \exp\left(-\frac{\beta}{2m}(p_x^2 + p_y^2 + p_z^2)\right). \tag{7.2.21}$$

But, the points (p_x, p_y, p_z) in momentum space can only have coordinates which are integral multiples of $(h/2a, h/2b, h/2c)$, i.e. they form a rectangular lattice. The density of these points will be $8abc/h^3$ and it follows that the number of molecules with momentum components in the ranges $(p_x, p_x + \Delta p_x)$, etc. (where Δp_x, etc. are supposed large compared with $h/2a$, etc.) should be taken to be

$$\frac{8V}{h^3} \Delta p_x \Delta p_y \Delta p_z \times \frac{1}{8} \zeta \exp\left(-\frac{\beta}{2m}(p_x^2 + p_y^2 + p_z^2)\right)$$

$$= N(2\pi mkT)^{-3/2} \Delta p_x \Delta p_y \Delta p_z \exp\left(-\frac{\beta}{2m}(p_x^2 + p_y^2 + p_z^2)\right), \quad (7.2.22)$$

using equation (7.2.17). This is Maxwell's classical result (3.7.3).

Since $\exp(-\beta\varepsilon_i) < 1$, equation (7.2.19) shows that $\bar{a}_i \ll 1$, i.e. very few of the available eigenstates are occupied at all. If, however, the gas density is so high and the temperature is so low that ζ can no longer be assumed small, the situation alters. First, note that ζ is bounded above, for $\bar{a}_i$ must be positive and hence $\zeta < e^{\beta\varepsilon_1}$, where ε_1 is the ground state for a molecule; putting $p = q = r = 1$ in equation (7.2.1) and taking $a = b = c = 1$m for simplicity, we find that $\varepsilon_1 = 3h^2/8m$; for hydrogen, therefore, $\beta\varepsilon_1 = 1.426 \times 10^{-17}/T$ and for any likely temperature we conclude that ζ will still be less than 1.

In cases where ζ cannot be regarded as small and the earlier analysis fails, we say the gas is *degenerate*. The gas is *strongly degenerate* if ζ is almost unity. Consider the extreme case $\zeta = 1$. Then

$$\bar{a}_j = 1/(e^{\beta\varepsilon_j} - 1) \quad (7.2.23)$$

and the lower energy levels will have very high occupation numbers, the lowest level having the largest number of all. Thus, as the temperature is lowered towards absolute zero, the molecules crowd into the lower energy states. This is referred to as *Einstein condensation*.

The argument of this section has validated the classical theory for an ideal gas of point molecules, provided the temperature is not too low or the density too high.

7.3 Gas of bosons with rotational energy

Next suppose the molecules of the boson gas possess energy components in addition to their kinetic energy of translation. Such components may be provided by their rotational motions about their mass centres or by vibrations of their constituent atoms relative to one another. In any case, we shall assume that the molecular Hamiltonian separates into a number of parts corresponding to the various energy sources, each involving coordinates and momenta not present in the others. The energy eigenfunction will then separate into corresponding factors and the energy eigenvalue into a sum of energies contributed by the various sources (as for equation (5.4.3)).

Let ε_i be the translational energy eigenvalues given by equation (7.2.1). For simplicity, consider the case where there is only one additional energy source and let η_r be its eigenvalues. Then the eigenvalues of the total molecular energy are given by $\varepsilon_i + \eta_r$, where i and r range independently over all positive integral values. The argument of section 7.1 is applicable after replacement of ε_i by $\varepsilon_i + \eta_r$ and multiplication or summation with respect to both subscripts in the crucial equations (7.1.5) and (7.1.6).

From equation (7.1.5) we deduce that

$$\ln \mathscr{Z} = -\sum_{i,r} \ln (1 - \zeta z_i w_r), \tag{7.3.1}$$

where $z_i = e^{-\beta \varepsilon_i}$ and $w_r = e^{-\beta \eta_r}$. Assuming ζ to be small, this equation can be expanded thus:

$$\ln \mathscr{Z} = \sum_{i,r,j} (\zeta z_i w_r)^j / j = \sum_j j^{-1} \zeta^j \left(\sum_i z_i^j \right) \left(\sum_r w_r^j \right), \tag{7.3.2}$$

where j is also summed over the range $(1, \infty)$.

Now suppose

$$Z(\beta) = \sum_i z_i = \sum_i e^{-\beta \varepsilon_i} \tag{7.3.3}$$

is the partition function associated with the molecule's energy component having eigenvalues ε_i. Approximating the discrete spectrum of eigenvalues by a continuous spectrum, as in the previous section, we calculate that

$$Z(\beta) = AV \int_0^\infty \varepsilon^{1/2} e^{-\beta \varepsilon} d\varepsilon = \tfrac{1}{2} AV \sqrt{(\pi/\beta^3)}, \tag{7.3.4}$$

using the identity (7.2.14). Then,

$$\sum_i z_i^j = \sum_i e^{-j \beta \varepsilon_i} = Z(j\beta). \tag{7.3.5}$$

Similarly, if $W(\beta)$ is the partition function for the other energy component, then

$$W(\beta) = \sum_r e^{-\beta \eta_r}, \tag{7.3.6}$$

and

$$\sum_r w_r^j = W(j\beta). \tag{7.3.7}$$

Equation (7.3.2) can now be written

$$\ln \mathscr{Z} = \sum_j j^{-1} Z(j\beta) W(j\beta) \zeta^j, \tag{7.3.8}$$

from which the thermodynamic state variables can be derived using formulae established in section 5.6.

The value of ζ is determined by equation (7.1.6), which can be expanded into the form

$$N = \sum_{i,r} \zeta z_i w_r (1 + \zeta z_i w_r + \zeta^2 z_i^2 w_r^2 + \ldots),$$

$$= \sum_j \zeta^j \left(\sum_i z_i^j \right) \left(\sum_r w_r^j \right),$$

$$= \sum_j Z(j\beta) W(j\beta) \zeta^j. \tag{7.3.9}$$

Using equation (5.6.26) (and remembering that $\zeta = e^{\mu\beta}$), substitution from equation (7.3.8) yields the following formula for the energy

$$U = -\sum_j \{ Z'(j\beta) W(j\beta) + Z(j\beta) W'(j\beta) \} \zeta^j. \tag{7.3.10}$$

Also, since $W(\beta)$ will be independent of the volume V and $Z(\beta)$ is proportional to V (equation (7.3.4)), it follows that $\ln \mathcal{Z}$ is proportional to V; hence, using equation (5.7.16), we obtain a formula for the pressure, viz.

$$P = \frac{1}{\beta} \frac{\partial}{\partial V}(\ln \mathcal{Z}) = \frac{1}{\beta V} \sum_j j^{-1} Z(j\beta) W(j\beta) \zeta^j. \tag{7.3.11}$$

If we now approximate to $O(\zeta)$, equation (7.3.9) shows that

$$\zeta = N / \{ Z(\beta) W(\beta) \} \tag{7.3.12}$$

and, then, equation (7.3.10) leads to

$$U = -N \frac{\partial}{\partial \beta} \{ \ln Z(\beta) + \ln W(\beta) \} \tag{7.3.13}$$

and equation (7.3.11) to

$$P = kNT / V, \tag{7.3.14}$$

which is the ideal gas law.

To calculate $W(\beta)$, suppose the molecule is diatomic and we take account of its rotational motion about the mass centre. The classical analysis for this type of gas has been performed in section 4.5. According to quantum theory (see e.g. L. D. Landau and E. M. Lifshitz, *Quantum Mechanics*, 2nd edn., pp. 101, 294, Pergamon, 1965) the energy eigenvalues for the rotation of a diatomic molecule are given by

$$\eta_r = \frac{h^2}{8\pi^2 C} r(r+1), \quad (r = 0, 1, 2, \ldots), \tag{7.3.15}$$

where $C = m_1 m_2 a^2/(m_1 + m_2)$ (m_1, m_2 are the masses of the two atoms and a is their distance apart). The rth eigenstate is $(2r + 1)$-fold degenerate (i.e. there are $(2r + 1)$ mutually orthogonal states with the same energy eigenvalues and these will contribute the same number of equal terms to the partition function). Hence

$$W(\beta) = \sum_{r=0}^{\infty} (2r+1) \exp\left(-\frac{h^2\beta}{8\pi^2 C} r(r+1) \right). \tag{7.3.16}$$

Provided β is not too large (i.e. $T \not\to 0$), it is reasonable to approximate this series by an integral, to get

$$W(\beta) = \int_0^{\infty} (2x+1) \exp\left\{ -\sigma x(x+1) \right\} dx, \tag{7.3.17}$$

where $\sigma = h^2\beta/8\pi^2 C$, and it then follows immediately that

$$W(\beta) = \frac{1}{\sigma} \left| \exp\left\{ -\sigma x(x+1) \right\} \right|_0^{\infty} = 1/\sigma. \tag{7.3.18}$$

Substituting for $Z(\beta)$ and $W(\beta)$ from equations (7.3.4) and (7.3.18) respectively in equation (7.3.13), we now find that

$$U = \frac{5}{2} kNT, \quad C_V = \partial U/\partial T = \frac{5}{2} kN, \tag{7.3.19}$$

in agreement with the classical result (4.5.9). More exactly, it may be proved (H. P. Mulholland, *Proc. Camb. Phil. Soc.*, **24**, 280, 1928) that

$$W = e^{\sigma/4} \left(\frac{1}{\sigma} + \frac{1}{12} + \frac{7}{480} \sigma + O(\sigma^2) \right) \tag{7.3.20}$$

from which we deduce that

$$C_V = \frac{5}{2} kN + \frac{Nh^4}{2880\,\pi^2 C^2 k} \cdot \frac{1}{T^2}, \tag{7.3.21}$$

showing how the classical value is approached as $T \to \infty$.

If, however, $T \to 0$, then $\beta \to \infty$ and equation (7.3.16) shows that $W \to 0$ and the contribution of the rotational energy dies away. As remarked in section 4.8, the classical theory is unable to explain this phenomenon, which is, however, in accordance with the experimental evidence.

7.4 Fermion gas

The line of argument pursued in section 7.1 for a gas of bosons will now be amended to apply to a gas of N indistinguishable fermions occupying a volume V.

For such particles, Pauli's exclusion principle limits the number a_i of molecules in each distinct energy state ε_i to be 0 or 1; thus, the summation in the grand partition function (7.1.3) must be carried out over all sets of integers $\{a_i\}$ obeying this restriction. Making the substitutions indicated by equations (7.1.4), we calculate that

$$\mathcal{Z} = \sum \zeta^{a_1 + a_2 + \cdots} z_1^{q_1} z_2^{q_2} \ldots ,$$

$$= (1 + \zeta z_1)(1 + \zeta z_2) \ldots ,$$

$$= \prod_{i=1}^{\infty} (1 + \zeta z_i). \tag{7.4.1}$$

Substituting in equation (5.6.25), we are led to the condition

$$\sum_i \frac{\zeta z_i}{1 + \zeta z_i} = N, \tag{7.4.2}$$

in replacement for the condition (7.1.6) applicable to bosons. Equation (7.1.8) yields similarly the replacement

$$\bar{a}_j = \frac{\zeta z_j}{1 + \zeta z_j} \tag{7.4.3}$$

for equation (7.1.9). It then follows that the mean energy is given by

$$U = \sum_i \frac{\varepsilon_i}{\zeta^{-1} \exp(\beta \varepsilon_i) + 1} \tag{7.4.4}$$

Note that the new equation (7.4.3), for the mean energy level occupation numbers, imposes no restriction on the maximum value of ζ. As $\zeta \to +\infty$, the gas enters a degenerate state in which $\bar{a}_j = 1$, except for very large values of j for which z_j is small. This means that all the lower energy levels will be occupied by exactly one fermion and that gaps will occur only at the higher levels.

7.5 Conduction electrons in metals

A major application of the statistical mechanics of a fermion gas is to those electrons belonging to the atoms of a metallic conductor which are free to move under the influence of an applied electric field and so to generate an electric current. Such electrons may be treated as being so weakly bound to their atoms, that they constitute an electron gas occupying the interstices between the atoms, the surface of the metal acting as the wall of a container—we assume that the P.E. of an electron within the metal is zero, but that this energy increases rapidly towards a high value if the electron moves to break the metal's surface, which accordingly imprisons the particle within its interior.

If we take the metal to be in the shape of a rectangular block of dimensions $a \times b \times c$, then the energy eigenvalues for an electron trapped within this region are given by equation (7.2.1). Since an electron has a pair of orthogonal spin states, each energy eigenstate gives rise to two equal terms in the partition function and associated sums; thus, equation (7.4.2) determining ζ takes the form

$$N = \sum_{p,q,r} \frac{2}{\zeta^{-1} e^{\beta \varepsilon} + 1}, \tag{7.5.1}$$

where ε is given by equation (7.2.1). Replacing the discrete spectrum of energy eigenvalues by a continuous spectrum for which the number of lines in the range $(\varepsilon, \varepsilon + d\varepsilon)$ is given at (7.2.4), we can replace the sum in the last condition by an integral to yield

$$N = 2AV \int_0^\infty \frac{\varepsilon^{1/2}\,d\varepsilon}{\exp \beta\,(\varepsilon - \mu) + 1}, \qquad (7.5.2)$$

where $\zeta = e^{\beta\mu}$.

The average number of electrons in the energy state ε_i is given by equation (7.4.3) to be

$$\bar{a}_i = 2/\{1 + \exp \beta(\varepsilon_i - \mu)\} \qquad (7.5.3)$$

(Note: The factor 2 allows for the two spin states associated with each energy eigenvalue.) Except for values of ε_i near to μ, the exponential term has either a very small or a very large value; thus $\bar{a}_i = 2$ for low energies and $\bar{a}_i = 0$ for high energies, showing that the electrons congregate in the lower energy states as far as this is permitted by the exclusion principle, and are only rarely found in states with energies greater than μ.

The number of electrons with energies in the range $(\varepsilon, \varepsilon + d\varepsilon)$ now follows from (7.2.4) and equation (7.5.3)—it is

$$dn = \frac{2AV\varepsilon^{1/2}\,d\varepsilon}{1 + \exp \beta(\varepsilon - \mu)}. \qquad (7.5.4)$$

For $T = 0\,\mathrm{K}$, $\beta = +\infty$ and so

$$\left.\begin{array}{ll} dn = 2\,AV\varepsilon^{1/2}\,d\varepsilon, & 0 < \varepsilon < \mu, \\ = 0 & \varepsilon > \mu, \end{array}\right\} \qquad (7.5.5)$$

Thus, μ is identified as the maximum energy an electron can attain at the absolute zero of temperature.

The mean energy of the electron gas follows from equation (7.5.4) in the form

$$U = 2AV \int_0^\infty \frac{\varepsilon^{3/2}\,d\varepsilon}{1 + \exp \beta(\varepsilon - \mu)}. \qquad (7.5.6)$$

For sufficiently low temperatures, $\beta\mu$ is large and the denominator of the integrands appearing in equations (7.5.2) and (7.5.6) is very nearly unity for values of ε in the range $(0, \mu)$; as ε increases outside this range, the denominator rapidly increases to take large values and the integrands then make little further contribution to the integrals. Hence, we can approximate these integrals by

$$\int_0^\mu \varepsilon^{1/2}\,d\varepsilon = \frac{2}{3}\mu^{3/2}, \qquad \int_0^\mu \varepsilon^{3/2}\,d\varepsilon = \frac{2}{5}\mu^{5/2}, \qquad (7.5.7)$$

respectively.

Equation (7.5.2) then yields the result

$$\mu = \left(\frac{3N}{4\,AV}\right)^{2/3} = \frac{h^2}{8m}\left(\frac{3\,N}{\pi V}\right)^{2/3}.$$ (7.5.8)

For copper, $N = 8.5 \times 10^{28}$ electrons per cubic metre and thus,

$$\beta\mu = 8.2 \times 10^4/T.$$ (7.5.9)

Hence, even for a temperature as high as 1000 K, $\beta\mu$ is about 80, and the approximation is good.

Using the approximation, equation (7.5.6) gives

$$U = \frac{4}{5}\,AV\mu^{5/2} = \frac{3}{5}\,N\mu.$$ (7.5.10)

We conclude that, to this order of approximation, U is independent of T and hence that the electron gas can make only a small contribution to the heat capacity of the metal. The situation is that most of the electrons cannot have their energies raised to a higher level, since such states are already occupied, and only the small proportion of electrons with energies in the neighbourhood of μ can be so affected. Thus, any heat energy communicated to the metal is largely consumed in raising the energy of vibration of the atoms forming the metal's crystal matrix and it is this component of energy alone which determines the heat capacity. This conclusion is confirmed by experiment, but could not be understood on the basis of the classical theory.

7.6 Photon gas

By treating the radiation as a gas of photons, our methods can be applied to the electromagnetic radiation in thermal equilibrium within an empty enclosure maintained at temperature T. Photons are bosons, but it is not necessary to determine the grand partition function and to use the argument of sections 7.1 and 7.2, since the number of photons is not prescribed and is treated as infinite. This is a consequence of the circumstance that each photon is associated with a normal mode of oscillation of the electromagnetic field at frequency v, the energy of the photon being given by the Einstein equation $\varepsilon = hv$. Since there is no upper limit to v, there is no limit on the number of photons and the occupation numbers a_i are not subject to constraint. Thus, there is no difficulty calculating the partition function for the whole closed system of radiation within the container and then applying the results of section 4.3.

If a_i denotes the number of photons in the energy state $\varepsilon_i = hv_i$, the corresponding energy state of the photon gas is given by

$$e_k = \sum_i a_i \varepsilon_i.$$ (7.6.1)

Hence, the partition function for the radiation is

$$Z = \sum e^{-\beta e_k} = \sum \exp\left(-\beta \sum_i a_i \varepsilon_i \right), \tag{7.6.2}$$

the outer summation being over all sets of positive integers $\{a_i\}$. Putting $z_i = e^{-\beta \varepsilon_i}$, this can be re-expressed thus:

$$Z = \sum z_1^{a_1} z_2^{a_2} \ldots = (1 + z_1 + z_1^2 + \ldots)(1 + z_2 + z_2^2 + \ldots) \ldots$$

$$= \prod_i \frac{1}{1 - z_i} = \prod_i (1 - e^{-\beta \varepsilon_i})^{-1}. \tag{7.6.3}$$

We can now use equation (4.3.15) to obtain the probability of finding the radiation in the energy state e_k, viz.

$$p_k = e^{-\beta e_k}/Z. \tag{7.6.4}$$

The mean occupation number for the ith photon state ε_i can now be written down as

$$\bar{a}_i = \frac{1}{Z} \sum a_i \exp\left(-\beta \sum_i a_i \varepsilon_i \right), \tag{7.6.5}$$

the summation being over all sets of positive integers $\{a_i\}$. Differentiating equation (7.6.2), we find

$$\frac{\partial Z}{\partial \varepsilon_i} = -\beta \sum a_i \exp\left(-\beta \sum_i a_i \varepsilon_i \right) = -\beta Z \bar{a}_i. \tag{7.6.6}$$

Hence,

$$\bar{a}_i = -\frac{1}{\beta} \frac{\partial}{\partial \varepsilon_i} (\ln Z). \tag{7.6.7}$$

Substitution from equation (7.6.3) now yields

$$\bar{a}_i = 1/(e^{\beta \varepsilon_i} - 1). \tag{7.6.8}$$

Although, for an enclosure of finite volume V, the spectrum of radiation frequencies is discrete, in practice the separation of the various lines is minute and it is possible to define a function $g(v)$ such that $g(v)\,dv$ gives the number of spectrum lines in the frequency range $(v, v + dv)$. If the enclosure is in the shape of a rectangular box, we have already calculated the number of energy eigenvalues in the range $(\varepsilon, \varepsilon + d\varepsilon)$ available to a boson trapped in the box, viz. the expression (7.2.4). If p is the magnitude of the particle's momentum, then $\varepsilon = p^2/2m$ (Note: non-relativistic quantum mechanics is always being assumed) and the number of eigenstates in the range $(p, p + dp)$ is found by substitution in (7.2.4) to be $4\pi V p^2\,dp/h^3$. For a photon, $p = hv/c$ and the number of eigenstates in the frequency range $(v, v + dv)$ is accordingly

$$g(v)\,dv = \frac{8\pi V}{c^3} v^2\,dv, \tag{7.6.9}$$

where we have introduced a factor 2 to allow for the circumstance that each photon has two spin states corresponding to orthogonal directions of polarization of the associated radiation. This result (7.6.9) can also be derived from the classical Maxwellian theory by determination of the normal modes of oscillation of the electromagnetic field (see R. H. Fowler, *Statistical Mechanics*, pp. 112–14, C.U.P., 1936).

Combining equations (7.6.8), (7.6.9), we now calculate that the mean number of photons with frequencies in the range $(v, v+\mathrm{d}v)$ is

$$\mathrm{d}n = \frac{8\pi V}{c^3} \cdot \frac{v^2 \, \mathrm{d}v}{e^{\beta h v} - 1}. \tag{7.6.10}$$

It then follows that the radiant energy for this frequency range is

$$\mathrm{d}U = hv \, \mathrm{d}n = \frac{8\pi V h}{c^3} \cdot \frac{v^3 \, \mathrm{d}v}{e^{\beta h v} - 1}. \tag{7.6.11}$$

This is *Planck's radiation law*. As $h \to 0$, $\mathrm{d}U \to 8\pi VkTv^2 \, \mathrm{d}v/c^3$, which is the classical *Rayleigh–Jeans law*.

The total radiant energy within the enclosure can now be calculated thus:

$$U = \frac{8\pi V h}{c^3} \int_0^\infty \frac{v^3 \, \mathrm{d}v}{e^{\beta h v} - 1} = \frac{8\pi V}{(ch)^3} (kT)^4 \int_0^\infty \frac{x^3 \, \mathrm{d}x}{e^x - 1}. \tag{7.6.12}$$

The definite integral can be evaluated in closed form by expansion as follows:

$$\int_0^\infty x^3 e^{-x} (1 - e^{-x})^{-1} \, \mathrm{d}x = \int_0^\infty \sum_{n=1}^\infty x^3 e^{-nx} \, \mathrm{d}x$$

$$= 6 \sum_{n=1}^\infty n^{-4} = \pi^4/15. \tag{7.6.13}$$

Hence

$$U = \frac{8\pi^5 k^4}{15(ch)^3} VT^4. \tag{7.6.14}$$

This is *Stefan's law*, already obtained at equation (2.6.8); however, the previously unknown constant a is now seen to be given by $a = 8\pi^5 k^4/15c^3 h^3$ $= 7.565 \times 10^{-16} \, \mathrm{J \, m^{-3} \, K^{-4}}$. (Note that the same calculation using the classical Rayleigh–Jeans law gives a divergent integral; quantum theory rectifies this anomaly.)

Equation (7.6.14) can also be derived from the partition function by application of equation (4.3.16). Combining equations (7.6.3) and (7.6.9), we find

$$\ln Z = -\int_0^\infty \frac{8\pi V}{c^3} v^2 \ln(1 - e^{-\beta h v}) \, \mathrm{d}v,$$

$$= -\frac{8\pi V}{(ch\beta)^3} \int_0^\infty x^2 \ln(1 - e^{-x}) \, \mathrm{d}x,$$

$$= \frac{8\pi V}{(ch\beta)^3} \int_0^\infty \left(\sum_{n=1}^\infty x^2 e^{-nx}/n \right) dx,$$

$$= \frac{8\pi^5}{45(ch\beta)^3} V. \tag{7.6.15}$$

Differentiation with respect to β now leads to equation (7.6.14). Further, differentiating $\ln Z$ with respect to V and applying equation (4.3.25), we calculate the radiation pressure to be

$$P = \frac{8\pi^5 k^4}{45 c^3 h^3} T^4 = U/3V; \tag{7.6.16}$$

i.e. the radiation pressure is one-third the energy density, as previously established at equation (2.6.3). Note that, for an ordinary gas of point bosons, we have proved (equation (7.2.12)) that, $P = 2U/3V = $ two-thirds the energy density.

Exercises 7

1. Show that the entropy of a boson gas is given by

$$S = \left(\frac{5}{3} U - \mu N \right) / T$$

and deduce that $G = \mu N$ (G is the Gibbs function).

2. Solve equation (7.2.7) for ζ to the second order of small quantities and deduce that, to the same order of approximation

$$U = \frac{3}{2} kNT - \frac{3}{32} \left(\frac{h^2}{\pi m} \right)^{3/2} \frac{N^2}{V} \frac{1}{\sqrt{(kT)}}.$$

3. If $\mathscr{Z}$ is the grand partition function for a gas of N electrons confined to a volume V, show that

$$\ln \mathscr{Z} = \frac{4}{3} AV\beta \int_0^\infty \frac{\varepsilon^{3/2} \, d\varepsilon}{\exp \beta(\varepsilon - \mu) + 1}$$

where μ is determined by equation (7.5.2). Using equation (5.6.26), deduce equation (7.5.6).

4. For the electron gas of the previous exercise, show that the entropy is given by

$$S = \left(\frac{5}{3} U - \mu N \right) / T.$$

For very low temperatures, using the result (7.5.12), show that $S = 0$. (Note: At these temperatures, all the low electron energy states are occupied and there is only one possible configuration for the particles to adopt; thus, $W = 1$ and Boltzmann's equation gives $S = 0$.)

5. Show that the pressure of the electron gas is given by

$$P = 2U/3V.$$

Deduce that, at low temperatures, $P(V/N)^{5/3} = $ constant.

6. For radiation in an enclosure of volume V at temperature T, show that the partition function Z is given by $\ln Z = U/(3kT)$. Deduce that the entropy is given by $S = 4aT^3/3$, where a is defined after equation (7.6.14). Deduce further that, for adiabatic expansion of the radiation, the following quantities are constant: VT^3, P^3V^4, PT^4, PV/T and U^3V.

Appendices

Appendix A: Electromagnetic field energy

If a charge distribution of density ρ, whose flow is specified by a current density vector $\mathbf{j}$, generates a field described by electric vectors $\mathbf{E}$ and $\mathbf{D}$, and by magnetic vectors $\mathbf{H}$ and $\mathbf{B}$, then using the SI system of units, Maxwell's equations take the form

$$\text{curl } \mathbf{E} = -\partial \mathbf{B}/\partial t, \quad \text{curl } \mathbf{H} = \mathbf{j} + \partial \mathbf{D}/\partial t. \tag{A.1}$$

It follows that

$$-\mathbf{E} \cdot \frac{\partial \mathbf{D}}{\partial t} - \mathbf{H} \cdot \frac{\partial \mathbf{B}}{\partial t} = \mathbf{H} \cdot \text{curl } \mathbf{E} - \mathbf{E} \cdot \text{curl } \mathbf{H} + \mathbf{E} \cdot \mathbf{j}. \tag{A.2}$$

Using the identity

$$\text{div}(\mathbf{E} \times \mathbf{H}) = \mathbf{H} \cdot \text{curl } \mathbf{E} - \mathbf{E} \cdot \text{curl } \mathbf{H}, \tag{A.3}$$

and integrating the previous equation over the interior of a sphere S of large radius R containing all the charges, we find

$$-\int \left(\mathbf{E} \cdot \frac{\partial \mathbf{D}}{\partial t} + \mathbf{H} \cdot \frac{\partial \mathbf{B}}{\partial t} \right) \mathrm{d}V = \int \text{div}(\mathbf{E} \times \mathbf{H}) \, \mathrm{d}V + \int \mathbf{E} \cdot \mathbf{j} \, \mathrm{d}V,$$

$$= \int (\mathbf{E} \times \mathbf{H})_n \, \mathrm{d}S + \int \mathbf{E} \cdot \mathbf{j} \, \mathrm{d}V, \tag{A.4}$$

having used the divergence theorem in the last step. Since $\mathbf{E}$ and $\mathbf{H}$ will be $O(1/R^2)$, the surface integral vanishes in the limit as $R \to \infty$. It follows that if, during a short time interval $\mathrm{d}t$, $\mathbf{D}$ and $\mathbf{B}$ increment by $\mathrm{d}\mathbf{D}$ and $\mathrm{d}\mathbf{B}$, then

$$-\int (\mathbf{E} \cdot \mathrm{d}\mathbf{D} + \mathbf{H} \cdot \mathrm{d}\mathbf{B}) \, \mathrm{d}V = \int (\mathbf{E} \cdot \mathbf{j} \, \mathrm{d}t) \, \mathrm{d}V. \tag{A.5}$$

According to the Lorentz formula, the force acting upon the charge $\rho \, \mathrm{d}V$ in the volume element $\mathrm{d}V$ is given by ($\mathbf{v}$ the velocity of flow)

$$\mathrm{d}\mathbf{F} = \rho \, \mathrm{d}V \, (\mathbf{E} + \mathbf{v} \times \mathbf{B}) \tag{A.6}$$

and the rate at which the field does work on this charge is accordingly

$$\mathrm{d}\mathbf{F} \cdot \mathbf{v} = \rho \, \mathrm{d}V \, \mathbf{E} \cdot \mathbf{v} = \mathbf{E} \cdot \mathbf{j} \, \mathrm{d}V, \tag{A.7}$$

since $\mathbf{j} = \rho \mathbf{v}$. It follows that the right-hand member of equation (A.5) gives the work done by the field on the whole charge distribution during the time dt. The field energy must be reduced by an equal amount to compensate and this reduction is given by the left-hand member of the equation.

We conclude, therefore, that if the field vectors $\mathbf{D}$ and $\mathbf{B}$ are incremented by $d\mathbf{D}$ and $d\mathbf{B}$, then the increase in the field energy is $(\mathbf{E} \cdot d\mathbf{D} + \mathbf{H} \cdot d\mathbf{B})$ per unit volume.

Appendix B: Stirling's formula

If n is a positive integer, it is easy to prove by repeated integration by parts that

$$n! = \int_0^\infty e^{-t} t^n \, dt. \tag{B.1}$$

Changing the variable of integration by the substitution $t = n(x + 1)$, it follows that

$$e^n n!/n^{n+1} = \int_{-1}^\infty \{(x+1)e^{-x}\}^n \, dx. \tag{B.2}$$

As x increases from -1 to 0, $(x+1)e^{-x}$ increases from 0 to 1; as x further increases from 0 to ∞, $(x+1)e^{-x}$ decreases from 1 to 0. Hence, if n is very large, $\{(x+1)e^{-x}\}^n$ is almost zero except in the vicinity of $x = 0$, where it takes values approaching unity; its graph is sketched below:

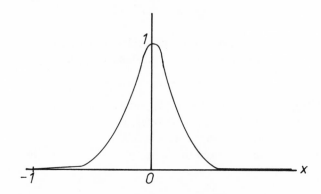

For values of x in the neighbourhood of zero,

$$(x+1)e^{-x} = 1 - \tfrac{1}{2}x^2 + O(x^3). \tag{B.3}$$

Thus, for such values, $e^{-nx^2/2}$ provides an approximation for $\{(x+1)e^{-x}\}^n$. However, if n is large, this approximation will also be good for other values of x. Equation (B.2) is accordingly approximated by

$$e^n n!/n^{n+1} \doteq \int_{-1}^\infty e^{-nx^2/2} \, dx \doteq \int_{-\infty}^\infty e^{-nx^2/2} \, dx = (2\pi/n)^{1/2}, \tag{B.4}$$

having used the identity (3.7.4).

Rearranging equation (B.4), we are led to Stirling's formula:

$$n! \doteqdot \sqrt{(2\pi)} n^{n+\frac{1}{2}} e^{-n}. \tag{B.5}$$

This formula is a good approximation even for low values of n. The values it gives for $n = 1(1)7$ are shown below, together with the percentage error in each case:

n	approx. $n!$	% error
1	0.922	7.8
2	1.919	4.1
3	5.836	2.7
4	23.506	2.1
5	118.019	1.7
6	710.078	1.4
7	4980.396	1.2

Appendix C: Evaluation of a contour integral

We require the value of the contour integral

$$\frac{1}{2\pi i} \oint z^{-U-1} Z^N \, dz \tag{C.1}$$

when N, U are very large positive integers related by the equations (3.5.3) and (3.5.4) (note that these imply $U \gg N$). Z is defined by the series

$$Z = z^{\varepsilon_1} + z^{\varepsilon_2} + \ldots, \tag{C.2}$$

where the indices ε_i are positive integers. We take the radius of convergence of this series to be $R(\leqslant 1)$ and the path of integration to lie within the circle of convergence and to embrace the point $z = 0$.

First consider the values taken by the integrand on the positive real axis, i.e. $z = x > 0$. Note that

$$\frac{d}{dx}(x^{-U-1} Z^N) = x^{-U-2} Z^N (NxZ'/Z - U - 1). \tag{C.3}$$

Writing, for convenience, $Z = \sum_{n=0}^{\infty} a_n x^n$, where the coefficients a_n are positive integers or zero. (N.B. the ε_i are not, necessarily, all different and, hence, the a_n can take integral values greater than 1.) Then

$$xZ'/Z = (\sum n a_n x^n)/(\sum a_n x^n), \tag{C.4}$$

and it follows that

$$\frac{d}{dx}\left(x\frac{Z'}{Z}\right) = \frac{(\sum a_n x^n)(\sum n^2 a_n x^{n-1}) - (\sum n a_n x^n)(\sum n a_n x^{n-1})}{(\sum a_n x^n)^2}. \tag{C.5}$$

Now

$$\left(\sum a_n x^n\right)\left(\sum n^2 a_n x^{n-1}\right) - \left(\sum n a_n x^n\right)\left(\sum n a_n x^{n-1}\right)$$

$$= \sum_{n,r} n^2 a_r a_n x^{r+n-1} - \sum_{n,r} rn a_r a_n x^{r+n-1}$$

$$= \frac{1}{2}\sum_{n,r}(r^2+n^2)a_r a_n x^{r+n-1} - \sum_{n,r} rn a_r a_n x^{r+n-1}$$

$$= \frac{1}{2}\sum_{n,r}(r-n)^2 a_r a_n x^{r+n-1} > 0. \tag{C.6}$$

Hence, xZ'/Z increases monotonically as x increases from 0 to R. It now follows from equation (C.3) that $x^{-U-1}Z^N$ first decreases and later increases as x increases from 0 to R (since $x^{-U-1}Z^N \to +\infty$ as $x \to 0$ and $x \to R$, this function cannot be monotonic for values of x between 0 and R). We conclude that the integrand has a unique minimum for some value of x between 0 and R satisfying the equation

$$-\frac{U+1}{x} + N\frac{Z'}{Z} = 0. \tag{C.7}$$

At this point x on the real axis, the derivative of $z^{-U-1}Z^N$ vanishes; hence $|z^{-U-1}Z^N|$ is stationary at this point. Moving from the point in either direction along the real axis, this modulus increases very rapidly, whereas it can be shown that for motion from the point parallel to the imaginary axis, the modulus rapidly decreases towards zero—i.e. the point is a col on the surface of values of $|z^{-U-1}Z^N|$. By taking the contour of integration to be a circle through the point x, having its centre at O, we can ensure that the modulus of the integrand is effectively zero on the whole contour, except in the immediate neighbourhood of x; i.e. only this small segment of the contour will contribute to the integral.

That this last statement is valid can be seen by comparing the value taken by the modulus of the integrand at any point $xe^{i\theta}$ on the circle with its value at x. The ratio is $\{|Z(xe^{i\theta})|/Z(x)\}^N$. Now

$$Z(x) = x^{\varepsilon_1} + x^{\varepsilon_2} + \ldots, \tag{C.8}$$

$$Z(xe^{i\theta}) = x^{\varepsilon_1}e^{i\varepsilon_1\theta} + x^{\varepsilon_2}e^{i\varepsilon_2\theta} + \ldots \tag{C.9}$$

All the terms in the first series (C.8) are positive and augment one another, whereas the exponential factors present in the second series (C.9) rotate the complex vectors representing x^{ε_1}, x^{ε_2}, etc. out of alignment and this reduces the modulus of their sum. (An exception would be if the integers ε_i had a common factor, i.e. $\varepsilon_i = n\eta_i$. Then, for $\theta = 2\pi/n, 4\pi/n$, etc. all the exponential factors would be unity again. In such a case, we would increase the magnitude of the unit of energy by a factor n and so eliminate this common factor.) We conclude that $|Z(xe^{i\theta})|/Z(x)$ is always less than unity at points on the contour other than x (i.e. $\theta = 0$) and that its Nth power, if N is very large, will be negligible.

At points on the circular contour near to x,

$$z = xe^{i\theta} = x(1 + i\theta - \tfrac{1}{2}\theta^2 + O(\theta^3)), \tag{C.10}$$

where θ is small. Putting $G(z) = \ln Z(z)$, we calculate that

$$\ln Z(xe^{i\theta}) = G(x + ix\theta - \tfrac{1}{2}x\theta^2 + O(\theta^3))$$
$$= G(x) + (ix\theta - \tfrac{1}{2}x\theta^2)G'(x) - \tfrac{1}{2}x^2\theta^2 G''(x) + O(\theta^3), \tag{C.11}$$

using Taylor's theorem. Hence, at $z = xe^{i\theta}$,

$$z^{-U-1} Z^N = x^{-U-1} \exp\{-(U+1)i\theta + N\ln Z\}$$
$$= x^{-U-1} \exp[-(U+1)i\theta + NG(x) + N(ix\theta - \tfrac{1}{2}x\theta^2)G'(x)$$
$$-\tfrac{1}{2}Nx^2\theta^2 G''(x) + O(\theta^3)]$$
$$= x^{-U-1} Z^N(x) \exp[-(U+1)i\theta + N(ix\theta - \tfrac{1}{2}x\theta^2)G'(x)$$
$$-\tfrac{1}{2}Nx^2\theta^2 G''(x) + O(\theta^3)]. \tag{C.12}$$

We now specify x and a function $g(x)$ by the equation

$$g(x) \equiv -\frac{U}{x} + NG'(x) = 0. \tag{C.13}$$

Since U is very large, the discrepancy between the equations (C.7) and (C.13) is not significant (if, however, the reader objects, we redefine x by this last equation). Then, differentiating, we find

$$g'(x) = \frac{U}{x^2} + NG''(x) = \frac{N}{x}G'(x) + NG''(x). \tag{C.14}$$

Equation (C.12) can now be written

$$z^{-U-1} Z^N = x^{-U-1} Z^N(x) \exp(-i\theta - \tfrac{1}{2}\theta^2 x^2 g'(x)). \tag{C.15}$$

Omitting the contribution of the whole of the circular contour except that of the small arc near to the point x, the integral (C.1) can now be approximated by

$$\frac{1}{2\pi} x^{-U} Z^N(x) \int \exp(-\tfrac{1}{2}\theta^2 x^2 g'(x)) \, d\theta, \tag{C.16}$$

where θ ranges over those small values for which the exponential integrand is not negligible. Clearly, as a final approximation, we can extend the range of integration to be $(-\infty, \infty)$ and use the identity (3.7.4) to yield our final result

$$\frac{1}{2\pi i} \oint z^{-U-1} Z^N \, dz = (2\pi g')^{-1/2} x^{-U-1} Z^N(x). \tag{C.17}$$

Bibliography

A selection of books which are recommended for further reading is given below:

Fowler, R. and Guggenheim, E. A., *Statistical Thermodynamics*, Cambridge University Press, 1960.

Gopal, E. S. R., *Statistical Mechanics and Properties of Matter*, Ellis Horwood, 1974.

Landau, L. D. and Lifshitz, E. M., *Statistical Physics*, Pergamon, 1980.

Landsberg, P. T., *Thermodynamics and Statistical Mechanics*, Oxford University Press, 1978.

Mandl, F., *Statistical Physics*, Wiley, 1971.

Pippard, A. B., *The Elements of Classical Thermodynamics*, Cambridge University Press, 1966.

Pointon, A. J., *Introduction to Statistical Physics*, Longman, 1975.

Rosser, W. G. V., *An Introduction to Statistical Physics*, Ellis Horwood, 1982.

Schrödinger, E., *Statistical Thermodynamics*, Cambridge University Press, 1964.

Tolman, R. C., *The Principles of Statistical Mechanics*, Oxford University Press, 1955.

Index